A Laboratory Manual for
Nonlinear Physics

with Maple for Scientists and Engineers

We wish to thank the members of our families for their continuing support, suggestions, tolerance, and their humoring of our fluctuating moods as the book and laboratory manual evolved from an idea to reality.

The Enns family	The McGuire family
Karen	Lynda
Russell	Colleen
Jennifer	Sheelo
Heather	Michael

Richard H. Enns   George McGuire

Laboratory Manual for Nonlinear Physics
with Maple for Scientists and Engineers

Birkhäuser
Boston • Basel • Berlin

Richard H. Enns   George McGuire
Department of Physics   Department of Physics
Simon Fraser University   University College of the Fraser Valley
Burnaby, BC V5A 1S6   Abbotsford, BC V2S 7M9
Canada   Canada

Library of Congress Cataloging In-Publication Data

Enns, Richard H.
   Nonlinear physics with Maple for scientists and engineers /
Richard H. Enns and George McGuire.
      p.   cm.
   Includes bibliographical references and index.
      ISBN 0-8176-3838-5 (alk. paper). -- ISBN 0-8176-3841-5 (lab
manual). -- ISBN 0-8176-3977-2 (set)
      1. Nonlinear theories--Data processing.   2. Mathematical physics-
-Data processing.   3. Maple (Computer file) I. McGuire, George,
1940-   . II. Title.
   QC20.7.N6E56   1997   96-40498
   530.15--DC21   CIP

Printed on acid-free paper
© 1997 Birkhäuser Boston   *Birkhäuser* 🄑®

ISBN 0-8176-3838-5 (Text)   ISBN 0-8176-3841-5 (Lab)   ISBN 0-8176-3977-2 (Set)
ISBN 3-7643-3838-5 (Text)   ISBN 3-7643-3841-5 (Lab)   ISBN 3-7643-3977-2 (Set)

Typeset by the authors in LaTex
Printed and bound by Hamilton Printing, Rensselaer, N.Y.
Printed in the U.S.A.

9 8 7 6 5 4 3 2 1

Contents

Preface

Science demands that all theory must be checked by experiment. Richard Feynman, Nobel Laureate in physics (1965), reminds us in a wonderful quote that "The test of all knowledge is experiment. Experiment is the sole judge of scientific truth."[1] It is because nonlinear physics can be so profoundly counterintuitive that these laboratory investigations are so important. This manual is designed to be used with the text *Nonlinear Physics with Maple for Scientists and Engineers*. Understanding is enhanced when experiments are used to check theory, so please attempt as many of the activities as you can. As you perform these activities, we hope that you will be amazed and startled by strange behavior, intrigued and terrorized by new ideas, and be able to amaze your friends as you relate your strange sightings! Remember that imagination is just as important as knowledge, so exercise yours whenever possible. But please be careful, as nonlinear activities can be addicting, can provide fond memories, and can awaken an interest that lasts a lifetime.

Although it has been said that a rose by any other name is still a rose, (with apologies to Shakespeare) the authors of this laboratory manual have, in an endeavor to encourage the use of these nonlinear investigations, called them experimental activities rather than experiments. A number of design innovations have been introduced:

A. Each of the included activities may be approached on three levels:

1. simplest—for theoreticians[2] and non-physicists—the emphasis is on observing and investigating the features of the nonlinear phenomena with the minimum of data gathering and analysis;

2. moderate—for physics majors and engineers—more emphasis on data gathering and analysis;

3. complex—activities designed for a stand-alone course for experimental physicists—a deeper and more profound analysis is required and modifications are suggested to stimulate ideas for research projects.

The authors have provided these three levels in an effort to permit instructors and students to choose the type of activity that best suits their needs. However,

---

[1]Feynman, Leighton, and Sands. *The Feynman Lectures on Physics*. Addison-Wesley, Reading, Mass., first edition, 1977, Volume 1, Page I-1.

[2]No insult intended—one of the authors of this manual is a theoretician.

a less rigorous and more relaxed approach than that normally used in upper-year undergraduate experiments is encouraged at all three levels.

B. Twenty-eight experimental activities are included to provide students and instructors with

1. a large number of permutations and combinations of paths through the manual's activities,

2. diverse investigations that

    (a) best suits their academic or personal interest,

    (b) have a duration which provides a chance for, and encourages, a successful conclusion of the activity,

    (c) are of a suitable level of complexity,

3. a choice of several simple activities or one complex research project.

Designing experimental activities that reward the students' effort with success and the pleasure of accomplishment is a major objective of this manual.

C. The experimental activities are directly related to the accompanying text's theory

1. to ensure that they are not perceived by the students as an unconnected, irrelevant, and meaningless time-consuming chore;

2. by placing a picture of a stopwatch (as shown here) in the margin of the text at the point the activity should be undertaken;

3. to ensure that they reinforce and provide physical examples of the text's material;

4. by using the text's Maple files to explore and confirm the experimental investigations.

The experimental activities are not assigned just for the sake of learning experimental techniques, but they are assigned to deepen and broaden the understanding of the nonlinear physics covered in the text.

D. The experimental activities are designed around the following principles:

1. they are simple and easy to perform;

2. they use apparatus that is commonly available in undergraduate laboratories;

3. they use apparatus that is easy to set-up;

4. the apparatus employed is kept as simple as possible in an effort not to confuse the understanding of the apparatus with the understanding of the physics;

5. some activities should be simple enough that they may be done as take-home projects;

6. they should be enlightening and enjoyable;

7. they contain a list of further investigations that probe more deeply into the physics involved;

8. they use Maple files, which are provided on an accompanying disk, to check that theory agrees with the observed physical behavior.

Most of the activities in this manual are of relatively short duration, have brief and uncomplicated theoretical explanations, and use apparatus that is relatively simple and easy to construct. However, the detailed interpretation of the physics may not be so easy or simple. So be prepared to bestir yourself.

Some of these activities are completely original, some were designed from theoretical discussions found in books, books that did not include any explanations of how to perform the activity or how to build the apparatus. Many of the activities are modifications of ideas found in diverse sources. The primary source for the borrowed ideas was the excellent journal *The American Journal of Physics* (AJP). We hope the citations and references give credit where credit is due.

We would like to offer some words of caution. First and foremost, we would like to say that although these experimental activities have been tested by the authors, they have not been student-tested under classroom conditions, but will be in January of 1997. This might mean that some theories and procedures, which are perfectly clear and lucid to us, will not in some instances be so clear and lucid to the students or instructors attempting these activities. Accordingly, the authors welcome any suggestions for improvement of the activities by

- giving a different interpretation of the physics,

- designing better or different apparatus,

- modifying or using a completely different procedure.

These are open–ended experiments; they are not meant to be cookbook laboratory activities. Please feel free to omit certain steps in the procedures, modify others, and just explore as your desires dictate.

As far as we know there is not a nonlinear physics laboratory manual on the market. The authors have done their best to reduce the number of ambiguities, typographical errors, and gaps in the procedures' steps. We hope that the number of egregious errors are few so that the experiments perform as stipulated and are rewarding and enjoyable.

Richard Enns and George McGuire

Acknowledgements
We would like to thank the following:

- The University College of the Fraser Valley for their help in producing this laboratory manual.

- David Cooke, a student at the University College of the Fraser Valley, for independently performing each experiment and offering suggestions for improvement.

- COREL DRAW for some of their computer sketches which we have taken the liberty of modifying.

Experimental Activity 1

Spin Toy Pendulum

Comment: This investigation is easy to perform and should not take more than 1 hour to complete.

Reference:

1. *Nonlinear Physics with Maple for Scientists and Engineers, 2.1.1.*

Object: To investigate a characteristic feature of nonlinear oscillatory motion, namely that the period varies with the amplitude of the oscillations.

Theory: One of the easiest ways of determining if an oscillating object is exhibiting nonlinear motion is to measure the period for a variety of different initial amplitudes. If the period varies with the amplitude, the restoring force is nonlinear. This activity uses a toy which, when oscillating, has its period increase as its amplitude increases. A simple pendulum behaves in this manner. The differential equation that describes the nonlinear oscillations of a pendulum

is
$$\ddot{\theta} + 2\gamma\dot{\theta} + \omega^2 \sin\theta = 0 \tag{1.1}$$

where θ is the angular displacement, γ is the linear damping coefficient, and ω the angular frequency.

This activity uses an inexpensive toy, sold under the name Revolution: The

spinning cylinder

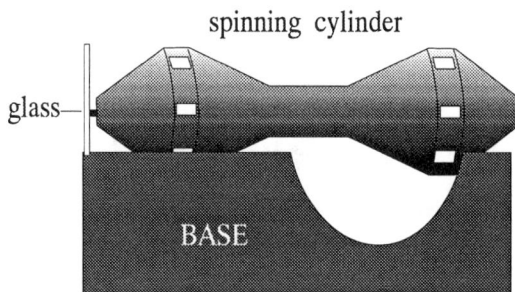

Figure 1.1: The spin toy pendulum.

World's Most Efficient Spinning Device[1], as a black box oscillator. Fig. 1.1 is a diagram of the device. The movable cylinder and base of this toy contain magnets which levitate and hold the spinning cylinder against the fixed glass end of the base. If the cylinder is given a sharp twist with the hand, it will continue to spin for a very long time. The toy has a low damping coefficient and therefore a very large quality (Q) factor. The quality factor is a dimensionless parameter that provides a measure of how fast a system dissipates energy. It is defined as $Q = \frac{2\pi E}{\Delta E}$ where E is the total energy and ΔE is the energy lost per cycle. For example, a quartz wrist watch has a $Q \approx 10^4$ whereas a piano string's Q is about 10 times lower.

If the cylinder is given a small initial twist it can be made to oscillate rather than rotate. For sufficiently small oscillations the simple harmonic oscillator equation prevails (Eq. (1.1) with $\sin\theta \approx \theta$) and $Q \approx \frac{\omega}{2\gamma}$ if the damping is not too large. In this limit, the period is given by $T = \frac{2\pi}{\omega}$ and is independent of the angular amplitude.

For larger oscillations the period begins to deviate from that predicted by the small oscillation formula and becomes noticeably amplitude-dependent. In this experimental activity, the period of the vibrations is measured against the angular amplitude and the mathematical relationship between these two quantities is investigated.

Procedure:

1. The spinning cylinder has white marks placed every 45° around its circumference. These marks can be used to estimate the initial amplitudes. (Newer versions of this toy may not contain these white markings so they may have to be painted on the cylinder by the experimenter.)

---

[1] Arbor Scientific, P.O. Box 2750, Ann Arbor, MI 48106-2750, phone 1-800-367-6695.

2. With a stop watch, and for an initial angular amplitude of less than 10°, measure the period of the resulting oscillations.

3. Using this small amplitude period (T), calculate the small amplitude frequency $f = \frac{1}{T}$, and small amplitude angular frequency $(\omega = \frac{2\pi}{T})$.

4. Measure the period for an initial amplitude of 22.5°. Can you detect any difference from the period measured previously?

5. Repeat the above Procedure for angles of approximately 45°, 67°, 90°, 112°, 135°, 157°, and for an angle as close to 180° as you can get.

6. Sketch a graph of the period as a function of the amplitude.

7. How do you know the restoring force is nonlinear?

8. For what initial amplitude is the nonlinearity first detectable?

9. Calculate an approximate low period damping coefficient (γ) for this oscillator. This can be done by measuring the length of time (t) it takes a small amplitude to decrease by a factor of two and then use the equation $\gamma = \frac{\ln(2)}{t}$.

10. Calculate the quality factor (Q) for small oscillations.

11. Use the provided Maple file X01SPIN.MWS to check whether the restoring force is similar to that of a simple pendulum. In the file, substitute your known values for γ and ω into the damped pendulum equation

$$\ddot{\theta} + 2\gamma\dot{\theta} + \omega^2 \sin\theta = 0. \tag{1.2}$$

For a given large initial amplitude, the Maple file will numerically solve and then make a plot of the angular displacement as a function of time. The file explains how the period may be found from the plot. Compare your experimental data for the period with that given by Maple.

Things to Investigate:

- Tilt the device so that the effect of gravity is reduced. Is the period dependent on the angle that the base is tipped?

- Investigate whether the damping term might be nonlinear.

- Use the Maple file X01BSTFT.MWS to see how accurately a polynomial can be used to approximate the curve shown on your plot of the period vs. angular displacement. Five or six pairs of data from the period vs. displacement graph are needed to run this file. Experiment with changing the order of the polynomial to produce a better fit.

Experimental Activity 2

Nonlinear Damping

Comment: This experiment is not difficult to do, but it will take about 3 hours to complete including the set-up time.

Reference:

1. *Nonlinear Physics with Maple for Scientists and Engineers, 2.1.3.*

Object: To investigate the dissipative force created by air drag.

Theory: The magnitude of the dissipative or drag force exerted on a body moving in a straight line through a viscous homogeneous medium depends upon a variety of parameters: the nature and density of the resisting medium as well as the size, shape, and speed of the moving object. To develop an expression for the drag force which correctly relates all these parameters can be quite difficult, especially over a large range in speed. This experiment provides a method of investigating nonlinear damping for low speeds.

Since the velocity is the only continuously changing variable, it is reasonable to assume the magnitude of the dissipative force on a body moving in a straight line through air will be a function of the instantaneous speed, i.e., $F(v)$. Even this simplification to one variable can still produce a complicated expression for the force. In an effort to simplify, aid understanding, and provide a workable approach to the problem, we assume that the expression for the dissipative force function is "well-behaved" over a low range of speeds and therefore can be expanded as a Maclaurin power series

$$F(v) = k_0 v^0 + k_1 v^1 + k_2 v^2 + k_3 v^3 \ldots \tag{2.1}$$

where the coefficients k_0, k_1, ... depend on the medium and physical parameters previously mentioned.

When the expression for the force is expressed as a power series it becomes possible to identify, pick out, and then use only one of the terms to model a particular type of drag. For example, in the case of sliding friction the drag force does not depend upon the speed, but only on the material making up the sliding surfaces and the normal force acting on the moving body. When the force produced by the sliding friction is much larger than the force produced by the speed dependent terms, the speed dependent terms may be ignored. Accordingly, only the first term of the above series is needed for sliding friction, so

$$F = k_0 v^0. \tag{2.2}$$

If the sliding object with a mass m is on a horizontal surface, the above term may be set equal to the standard expression for the force of sliding friction ($F = -\mu m g$) so

$$k_0 = -\mu m g. \tag{2.3}$$

Here μ is the coefficient of sliding friction and g is the acceleration due to gravity.

In situations such as a glider moving on a nearly frictionless airtrack the coefficient of sliding friction is very small and the speed dependent terms dominate. If the airtrack glider is moving slowly through the air and if it produces very little air turbulence, the magnitude of the dissipative force usually varies as a function of the first power of the instantaneous speed (v) so

$$F = k_1 v^1. \tag{2.4}$$

This assumes that all the other speed-dependent terms are of less importance. This type of dissipative force is known as a Stokes or a laminar damping force.

If the moving airtrack glider produces a lot of turbulence, e.g., due to a higher speed or to its shape, the magnitude of the dissipative force may be more accurately modeled by the term containing the square of the instantaneous speed

$$F = k_2 v^2. \tag{2.5}$$

This type of dissipative force is known as Newtonian damping.

In this activity the instantaneous speeds and dissipative forces are measured for a variety of speeds and then a $\ln(F)$ versus $\ln(v)$ graph is drawn. The values for n and k are found by graphical analysis. After the values for the exponent n and coefficient k are known it might be possible to identify which of the power series terms best fits the experimental results.

In this experiment, a moving air track glider passes between two photocell timers as shown in Fig. 2.1. The net force on the glider can be calculated by first

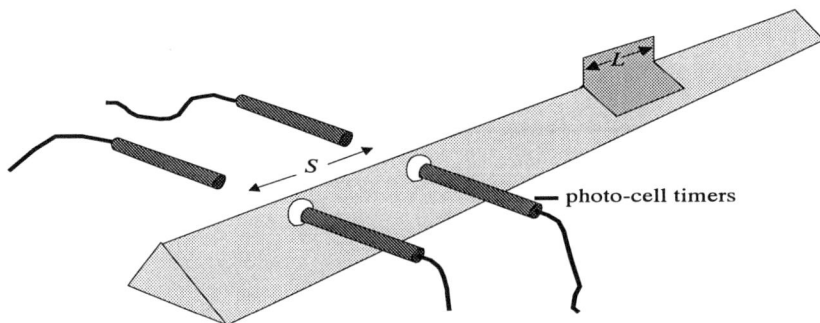

Figure 2.1: An airtrack glider moving between two photocell timers.

measuring the times (t_1, t_2) it takes the glider to pass by the two photocells.

Then the speeds (v_1, v_2) at these two locations are calculated by measuring the length of the glider (L), and dividing by the time the glider takes to pass by each timer, thus,

$$v_1 = \frac{L}{t_1}, \quad v_2 = \frac{L}{t_2}. \tag{2.6}$$

The dissipative force (F) acting on the glider is found by equating the glider's change in kinetic energy to the work done on the glider as it moved the distance s between the timers:

$$W = Fs. \tag{2.7}$$

Since the change in the glider's kinetic energy is equal to the work

$$Fs = \frac{mv_2^2}{2} - \frac{mv_1^2}{2}, \tag{2.8}$$

the magnitude of the dissipative force is equal to

$$F = \left| \frac{mv_2^2 - mv_1^2}{2s} \right|. \tag{2.9}$$

If the distance (s) between the timers is kept small, then the two measured speeds (v_1, v_2) will be approximately equal. The speed (v) which creates the dissipative force is close to the average of the two calculated speeds. The average speed $(\bar{v})$ is calculated by using

$$\bar{v} = \frac{v_2 + v_1}{2}. \tag{2.10}$$

Once a number of different forces and their related average speeds have been calculated, the mathematical relationship that unites them is determined.

Procedure:

1. Set up the apparatus as shown in Fig. 2.2.

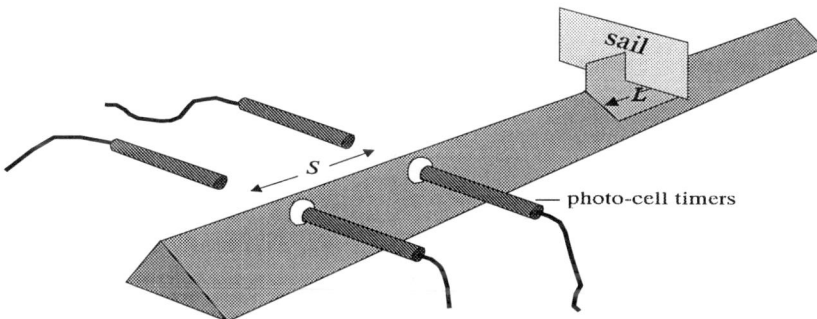

Figure 2.2: Apparatus to investigate force as a function of speed.

2. Measure the mass m of the glider.

3. The distance s should be about 20 cm.

4. The air track should be level.

5. Use a glider that has a length (L) of about 15 cm and a cardboard sail that has an area between 100 and 200 cm^2. The sail creates the majority of the air drag.

6. Launch the glider and record the times (t_1, t_2) it takes to pass by the timers. Using Eq. (2.6), calculate the speeds (v_1, v_2).

7. Calculate the dissipative force using Eq. (2.9).

8. Calculate the average velocity ($\bar{v}$) using Eq. (2.10).

9. Repeat steps 6,7, and 8 for at least six widely different speeds.

10. Draw a graph of $\ln(F)$ versus $\ln(\bar{v})$.

11. Analyze the graph to find the mathematical relationship between F and $\bar{v}$. Determine the values for the exponent n and the proportionality constant k.

12. The experimental value for n may not be an integer. If you were to model your experimental result by using only a single term from the Maclaurin series (Eq. 2.1), what integer would you use for n? Give reasons why the experimental value for n is different from an integer value.

13. Place the experimental data for v and F into the provided Maple file X02BSTFT.MWS to see if the Maple file can produce a power series ($y \equiv F$, $x \equiv v$) of the form $y = ax^2 + bx$ or even $y = ax^3 + bx^2 + cx$. Which of the power series best represents your data? With these new equations use the text's Maple file MF02 to numerically solve the equations and then compare the results with the experimental data.

Things to Investigate:

- Use sails of different shapes, for example, wedges or cones, to see how they modify the experimental values for the exponent n and/or coefficient k.

- Modify this activity by using an electric fan to blow across the sail to create the drag force and a spring attached to the glider to measure the force on the stationary glider. Compare the values for n and k with the values found in the above procedure.

- This activity might be modified to measure the drag on larger objects, such as a person coasting on bicycle.

Experimental Activity 3

Anharmonic Potential

Comment: This activity can be done in under three hours, if the apparatus has been previously set up.

Reference:

1. *Nonlinear Physics with Maple for Scientists and Engineers, 2.1.4.*

Object: To produce an anharmonic potential, determine its analytic form, and investigate the period of oscillations governed by this potential.

Theory: Anharmonic oscillators have nonlinear restoring forces, so unlike linear oscillators, their periods depend on the magnitude of their amplitudes. The motion of the anharmonic oscillator in this experimental activity is controlled by a potential energy function $(U(x))$ that consists of two terms. One term is the magnetic potential energy $(U_m(x) = \frac{b}{x^n})$ and the other term is the gravitational potential energy $(U_g(x) = cx)$, so the total potential energy is

$$U(x) = \frac{b}{x^n} + cx, \tag{3.1}$$

where b, c, and n are all positive constants. A representative plot of the above function is illustrated in Fig 3.1. Examination of the plot shows that oscillations

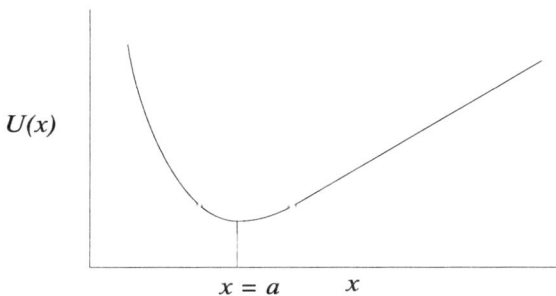

U(x)

x = a   *x*

Figure 3.1: Potential energy (U) vs. position (x).

can occur around the equilibrium point $(x = a)$. For small oscillations about the equilibrium point, Eq. (3.1) can be Taylor expanded to yield

$$U(x) = U(a) + U'(a)(x - a) + \frac{1}{2!}U''(a)(x - a)^2 + \frac{1}{3!}U'''(a)(x - a)^3 + ... \quad (3.2)$$

Since the slope of the potential energy curve is equal to zero at the equilibrium point $(x = a)$

$$U'(a) = -\frac{nb}{a^{n+1}} + c = 0, \quad (3.3)$$

the location of the equilibrium point is

$$a = \left(\frac{nb}{c}\right)^{\frac{1}{(n+1)}}. \quad (3.4)$$

Ignoring the terms beyond the quadratic we have

$$U(x) = U(a) + \frac{1}{2!}U''(a)(x - a)^2, \quad (3.5)$$

and since the force $F = -\frac{dU}{dx}$,

$$F = -k(x - a), \quad (3.6)$$

where

$$k = U''(a) = \frac{n(n + 1)b}{a^{(n+2)}}. \quad (3.7)$$

This approximation gives the (simple) harmonic limit. In this case, the restoring force is linear in the displacement with k playing the role of the spring constant. For larger displacements the higher order terms in the Taylor expansion must be considered. The restoring force is no longer linear and the potential is referred to as anharmonic.

Assuming that no dissipative forces are acting, the net force on a mass in the anharmonic potential given by Eq. (3.1) is found using $F = -\frac{dU}{dx}$, which gives

$$F = \frac{nb}{x^{(n+1)}} - c. \quad (3.8)$$

Since the net force on the mass is $F = ma$ or equivalently $F = m\ddot{x}$, the differential equation describing the motion is

$$m\ddot{x} - \frac{nb}{x^{(n+1)}} + c = 0. \quad (3.9)$$

In general, analytical solutions of Eq. (3.9) are not possible, so numerical methods are usually applied. This activity will use the Maple file X03AHR.MWS to check the experimental behavior of the oscillator governed by the anharmonic potential.

In this experimental activity, the values for b, c, and n are determined, and then substituted into Eq. (3.9). If the initial values for position and speed are also given, a numerical solution for position as a function of time is easily found using Maple. The period for large amplitude oscillations is determined from this numerical solution.

Procedure:

1. Set up the apparatus as shown in Fig. 3.2 using a 1.2 m airtrack. Make sure the air track is level.

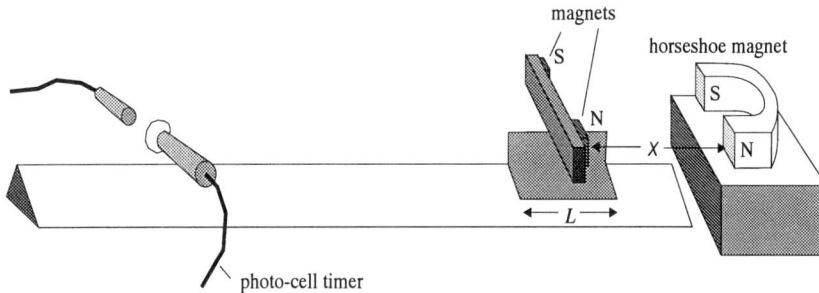

Figure 3.2: Apparatus to determine U_m as a function of x.

2. Mount the gated photocell timer approximately 0.60 m from the securely fastened powerful horseshoe magnet. If a horseshoe magnet is not available, two strong neodymium disk magnets fastened to a steel bar should work as well.

3. Attach two small neodymium disk magnets to the light horizontal plastic bar which is fastened to the top of the glider. The poles of the small magnets should be arranged so that a repulsive force exists between the horseshoe magnet and the glider's small magnets.

4. Measure the mass of the glider (magnets attached).

5. Place the glider near the large repelling magnet. For the first run, the separation between the repelling magnets should be about 0.050 m.

6. Release the glider, and record the time (t) it takes to pass by the photocell at the other end of the track.

7. Calculate the glider's speed ($v = L/t$) as it passes by the timer. (In theory the glider should keep picking up speed as it moves further away from the repelling magnet, but after it is about 0.50 m from the magnet, the repelling force becomes negligible, and the glider maintains a constant speed.)

8. Calculate the final kinetic energy ($K = \frac{mv^2}{2}$) of the glider. The final kinetic energy is actually equal to the change in kinetic energy, because the glider starts from rest. The change in kinetic energy is equal to the negative change in magnetic potential energy. The magnetic potential energy (U_m) is assumed to be zero when the distance between glider and the magnet is large, so the initial magnetic potential energy just equals the final kinetic energy of the glider.

9. Record the initial position (x) of the glider and its magnetic potential energy (U_m).

10. Repeat the above procedure for at least six different initial values of x from 0.05 to 0.20 m.

11. Record the data for x and U_m in an appropriate table.

12. Using the data in the table, find the mathematical relationship between U_m and x. The equation should have the form

$$U_m(x) = \frac{b}{x^n}.$$

The values for b and n can be found by plotting and analyzing a $\ln(U_m)$ vs. $\ln(x)$ graph. (A calculator with a linear regression program may be used to avoid making the plot.)

13. Now incline the air track as shown Fig. 3.3. The wood block should have a height of 2 to 3 cm for a 1.2 m long air track. Note: The height of the wood block was chosen to make the maximum gravitational potential roughly match the magnetic potential energy produced by the authors' magnet at $x = 0.10$ m.

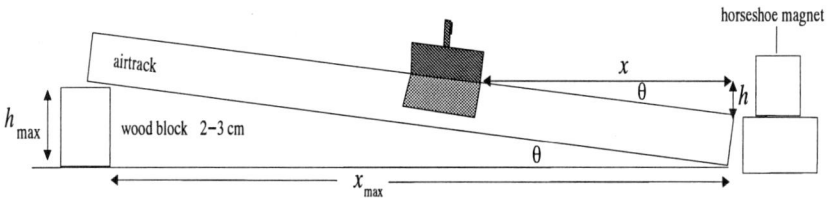

Figure 3.3: Inclined airtrack.

14. Calculate the angle of elevation $(\tan\theta = \frac{h_{max}}{x_{max}})$.

15. The gravitational potential energy of the glider at any position on the air track is $U_g = mgh$ where $h = x\tan\theta$ so $U_g = mgx\tan\theta$.

16. The value for c in Eq. (3.1) is equal to $mg\tan\theta$. The values b, c, and n are now all known.

17. Use Eq. (3.4) to calculate where the theoretical equilibrium point $(x = a)$ is located.

18. Place the glider back on the inclined track and locate the experimental equilibrium point.

19. Compare the theoretical and experimental values for the equilibrium point.

20. Displace the glider a small distance (≈ 0.01 m) from the equilibrium point, release the glider and measure the small oscillation period of the resulting vibrations.

21. Calculate the value for the spring constant using Eq. (3.7). After the value for k is known, calculate the small oscillation period by using

$$T = 2\pi\sqrt{\frac{m}{k}}.$$

22. Displace the glider a large distance from the equilibrium point. Record the distance and release the glider. Measure the period of the resulting oscillations.

23. Repeat the above procedure for a different large amplitude.

24. Use the Maple file X03AHR.MWS to check if the measured values for the large amplitude period agree with the calculated (numerical) periods.

Things to Investigate:

- Check if the measured values for the small oscillation periods can be used to modify the values of the constants b and n so that the periods predicted by the equation agree more accurately with the measured values.

- In the provided Maple file X03AHR.MWS change the values of b or n to make the theoretical large amplitude periods conform more accurately to the measured values for the periods.

- Change the model (equation) to account for dissipative forces (sliding or air friction). Alter the Maple file so that the predicted amplitudes of the damped vibrations agree with the experimental values.

- Place the whole apparatus on a wheeled platform that can be vibrated back and forth by a strong motor. This motion can be modeled by introducing a sinusoidal driving force which should permit an investigation of possible chaotic motion.

Experimental Activity 4

Iron Core Inductor

Comment: This investigation should not take more than 2 hours to complete.

Reference:

   1. *Nonlinear Physics with Maple for Scientists and Engineers, 2.3.1.*

Object: To produce current oscillations in a nonlinear tank circuit and to determine the period of these oscillations as a function of the current amplitude.

Theory: One of the most easily identified characteristics of all nonlinear oscillators is that the period varies with the amplitude of the oscillations. This activity attempts to illustrate and verify this behavior. The nonlinear oscillations are created using the nonlinear tank circuit shown in Fig. 4.1. The circuit consists of a high voltage capacitor (C) connected to an iron core inductor. It

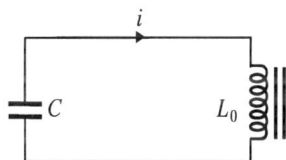

Figure 4.1: A nonlinear tank circuit.

is the iron core of the inductor that makes the circuit behave in a nonlinear manner.

   To understand how the iron core inductor introduces the circuit's nonlinearity, assume that the inductor is connected to a current source. As the electric current from the power source increases, the current through the inductor increases, and at first the strength of magnetic flux keeps pace. However, as the current increases even further, the linearity between current and flux no longer holds. The reason for this loss of linearity is that the magnetic field in the inductor's iron core begins to approach its magnetic saturation limit. Further increases in current produce very small increases in the flux. A plot of the inductor's current i vs. the inductor's flux ϕ should look something like that given in Fig. 4.2. The equation that models the current–flux relationship for

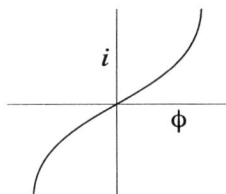

Figure 4.2: Current (i) vs. flux (ϕ) for an iron core inductor.

this circuit is given by

$$i = \frac{N\phi}{L_0} + A\phi^3, \tag{4.1}$$

where L_0 represents the self-inductance with the core present, N is the number of turns in the coil, and A is a constant dependent on the composition and construction of the core.

Using the circuit shown in Fig. 4.1, assume that at $t = 0$ seconds the capacitor (C) is fully charged and no current is flowing. As the capacitor begins to discharge, a time varying current i flows in the tank circuit. The potential drops must sum to zero, so

$$\frac{q}{C} + N\frac{d\phi}{dt} = 0. \tag{4.2}$$

Differentiating Eq. (4.2) to find the current gives

$$i = -NC\frac{d^2\phi}{dt^2} \tag{4.3}$$

and if Eq. (4.3) and Eq. (4.1) are combined they produce

$$\ddot{\phi} + \alpha\phi + \beta\phi^3 = 0, \tag{4.4}$$

where $\alpha = \frac{1}{L_0 C}$ and $\beta = \frac{A}{NC}$. The cubic term ($\beta\phi^3$) makes this a nonlinear equation. Eq. (4.4) is an undamped, homogeneous, "hard" spring differential equation. It has an analytical solution which can be expressed in terms of an elliptic function (sn or cn)[1]. For a hard spring oscillator, the period of the oscillations decreases with increasing amplitude.

Since the flux ϕ is difficult to measure experimentally, this activity measures the current vs. time rather than flux vs. time. The relationship between current and flux is given by Eq. (4.1). The activity checks the validity of this theory.

Procedure:

Be careful performing this activity as high voltages are present. Voltages such as these can produce severe electrical shocks. Do not touch the charged capacitor with your fingers and always make sure the capacitor is completely discharged before adjusting the circuit.

---

[1]See Section 4.2.4 of the accompanying text.

1. Wire the circuit shown in Figure 4.3. Use a demonstration transformer for

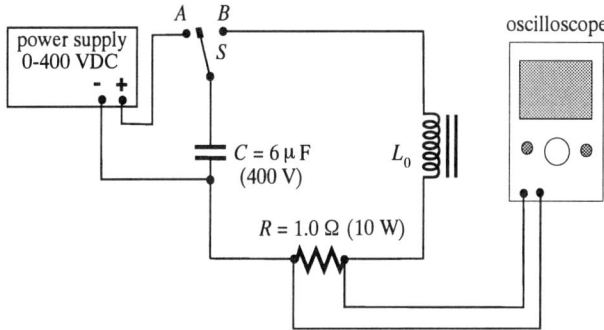

Figure 4.3: The apparatus.

the iron core inductor. Demonstration transformers come with a variety of interchangeable low resistance coils and have a removable or adjustable section of the core. This permits a large number of permutations and combinations of field strengths and inductances to be explored. (The transformer used in this experiment was purchased from CENCO and was listed in their 1995 catalogue as the Modular Transformer System, U-core and Yoke #56211T. The coil we used had a stated inductance of 36 mH (no core) and with the core present, $L_0 \approx 1.0$–4.0 H. The inductor had a resistance of 9.5 Ω and consisted of 1000 turns of wire.) The 1.0 Ω resistor is used to sample the signal and since its resistance is 1.0 Ω, the value of the potential drop measured across the resistor equals the value of the current flowing through it. The total resistance of the circuit, due mainly to the inductor, damps the oscillations and permits the period vs. amplitude dependence to be found from a single trace on the cathode ray tube (CRT). Set the movable part of the iron core as shown in Fig. 4.4. A storage oscilloscope permits the trace to be retained and studied in a

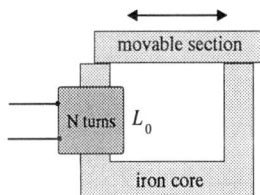

Figure 4.4: The transformer used as the iron core inductor.

more detailed and leisurely manner.

2. With the switch at A, charge the capacitor to 5 volts, and then move the switch to position B. Measure the resulting trace's period (T_0) and calculate the angular frequency. Using this angular frequency calculate the value for the circuit's inductance L_0 by using $\omega = \frac{1}{\sqrt{L_0 C}}$.

3. With the power supply voltage set to 50 volts, charge the capacitor and then move the switch from A to B. By observing the amplitude and period of the oscillations, determine whether the decaying oscillations are non-linear. You might wish to use the 10x beam expander on the oscilloscope to make the trace's pattern more discernible.

4. Repeat for larger and larger voltages. Increasing the voltage in steps of 50 volts is suggested.

5. At what voltage do you first see identifiable nonlinear behavior? The shape of the current vs. time trace on the CRT should be similar to the one shown in Figure 4.5.

Figure 4.5: Current vs. time trace.

6. If the signal decays too rapidly, try adjusting the movable part of the core to change the size of the overlap region.

7. Examine the CRT trace to confirm that as the amplitude decreases the period increases.

8. Using the oscilloscope trace, record the values for the decreasing amplitude and the lengthening period.

9. Make a graph of the period (T) vs. the current amplitude (i_0). What is the mathematical relationship between the period and amplitude?

10. Maple file X04TANK.MWS produces a plot similar to that shown in Fig. 4.5. Does the same period–amplitude relationship hold for the Maple plot as that found experimentally?

11. Try to find the approximate experimental values that can be substituted into this Maple file so that they model your results. Do these results confirm or repudiate the nonlinear model given in the theory?

Things to Investigate:

• Repeat the activity using a larger capacitor, e.g., 10 μF. Make sure the voltage rating of the new capacitor is at least 400 V.

• For a set power supply voltage, explore how the CRT signal changes as the core's flux linkage is changed. Alter the flux linkage by changing the position of the movable core. Start the movable core in a position that produces a small core overlap, and then increase the overlap in small steps. Which position gives the lowest damping?

Experimental Activity 5

Nonlinear LRC Circuit

Comment: This is an easy investigation to perform. It should not take more than 2 hours to complete.

References:

1. *Nonlinear Physics with Maple for Scientists and Engineers, 2.3.1.*

2. [FJB85] This article contains circuits which model both the soft and hard spring oscillators.

Object: To produce a piecewise linear (nonlinear) LRC circuit that models a "soft" spring oscillator and to investigate the relationship between the period and the amplitude of the oscillator.

Theory: A 1-dimensional mechanical soft spring oscillator has a restoring force whose magnitude is given by the expression

$$F = ax - bx^3 \tag{5.1}$$

where a and b are positive constants and x is the displacement of the vibrating object from its equilibrium position. If damping is included the differential equation describing the object's motion is

$$m\ddot{x} + c\dot{x} + ax - bx^3 = 0 \tag{5.2}$$

where c is the dissipative force constant. This equation may be transformed into the more standard and convenient form

$$\ddot{x} + 2\gamma\dot{x} + \alpha x - \beta x^3 = 0, \tag{5.3}$$

where $\gamma = \frac{c}{2m}$, $\alpha = \frac{a}{m}$, and $\beta = \frac{b}{m}$.

Oscillations produced by "soft" springs have periods that increase as the amplitude increases. It is this characteristic that will be used to identify the presence of "soft" spring oscillations in the electrical circuit used in this activity.

An electrical circuit that has its oscillations governed by an analogous equation to that of Eq. (5.3) is shown in Fig. 5.1. To understand how the circuit

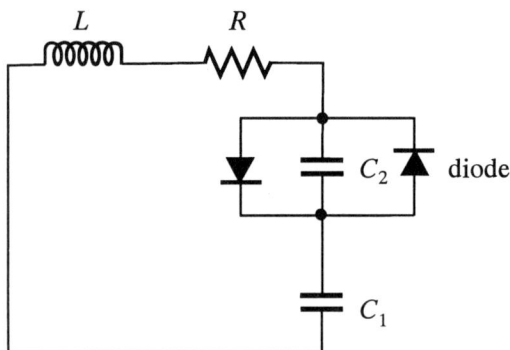

Figure 5.1: Circuit used to produce a soft spring characteristic.

functions, we use Kirchhoff's voltage rule that says that the algebraic sum of the potential drops around the closed loop must sum to zero. Therefore

$$V_L + V_R + V_C = 0 \qquad (5.4)$$

where V_L is the potential drop across the inductor, V_R is the voltage drop across the resistor, and V_C is the voltage drop across both capacitors. Eq. (5.4) is equivalent to

$$L\ddot{q} + R\dot{q} + V_C(q) = 0 \qquad (5.5)$$

where q is the electric charge. In the above equation the value for V_C will vary according to the amount of charge on the capacitors C_1 and C_2. A plot of the voltage across both capacitors as a function of the charge on the capacitors is as shown in Fig. 5.2. The value of the slope of each linear region is equal to the

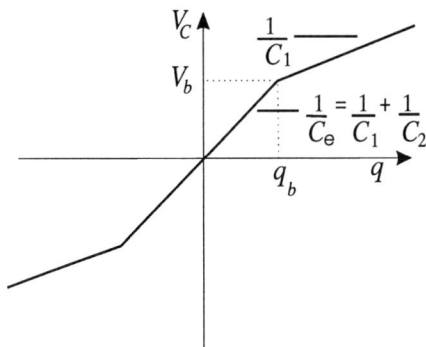

Figure 5.2: Voltage vs. charge across both capacitors.

reciprocal of the capacitance in that region. The reason for this piecewise linear shape can readily be understood by noting the diodes placed across C_2. When the potential drop across C_2 is lower than a certain voltage—about 0.70 volts for silicon diodes—the diodes do not conduct. In this instance, the diodes act as open switches and so the circuit's equivalent capacitance (C_e) is that of the two capacitors in series so

$$\frac{1}{C_e} = \frac{1}{C_1} + \frac{1}{C_2}. \tag{5.6}$$

As the charge on the capacitors increases, the potential drop across C_2 will reach 0.7 volts and the diodes begin to conduct. As long as the voltage remains higher than 0.7 volts the diodes act as closed switches and effectively remove the capacitor C_2 from the circuit. The circuit's equivalent capacitance increases or the slope of the line shown in Fig. 5.2 decreases to the value of $\frac{1}{C_1}$. Accordingly, the straight line bends downward at $V_b \approx 0.70$ volts. Remember, the slope is the reciprocal of the capacitance. The shape of the piecewise plot shown in Fig. 5.2 has now been explained. In Eq. (5.5), the form of $V_c(q)$ is given by the following piecewise function:

$$V_c = q \begin{cases} C_e^{-1} & |q(t)| < q_b \\ C_1^{-1} & \text{otherwise} \end{cases}. \tag{5.7}$$

To show how the piecewise function in Fig. 5.2 might produce behavior similar to that of a "soft" spring, the three straight line segments are approximated by a continuous function of the form

$$V_C = aq - bq^3. \tag{5.8}$$

Plotting Eq. (5.8) gives Fig 5.3 which has a shape similar to Fig. 5.2 for a certain

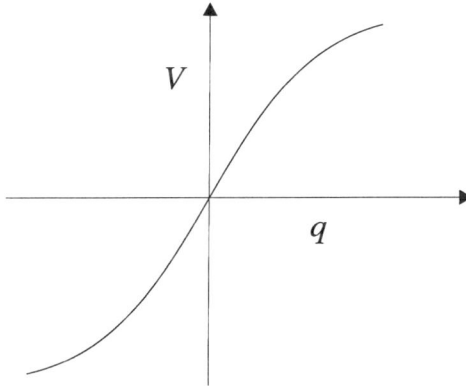

Figure 5.3: Voltage as function of charge.

limited range of values for the charge q. The reason for this restriction is that if Eq. (5.8) is plotted over a large range of q, the curve would turn over at large $|q|$. The piecewise function never does.

Substituting Eq. (5.8) into Eq. (5.5) gives the differential equation for a soft spring,

$$L\ddot{q} + R\dot{q} + aq - bq^3 = 0. \tag{5.9}$$

Writing the above equation in the more traditional form gives,

$$\ddot{q} + 2\gamma\dot{q} + \alpha q - \beta q^3 = 0, \tag{5.10}$$

where $\alpha = \frac{a}{L}$ and $\beta = \frac{b}{L}$. This equation shows that the circuit used in this activity is approximately that of a "soft" spring oscillator.

Procedure:

1. Construct the circuit shown in Figure 5.4. The circuit's inductor should

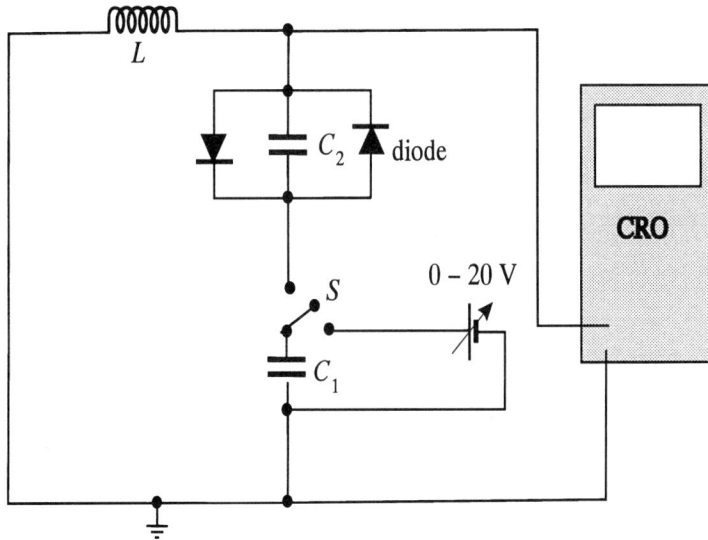

Figure 5.4: Soft spring circuit.

have an inductance between 0.80 H, (the Berkeley solenoid) and 0.0040 H (PSSC solenoid). Perform the initial run with two equal 0.47 μF capacitors. The resistance R shown in Fig. 5.1 is provided by the resistance of the wires, diodes, and the solenoid. This resistance is what damps the oscillations.

2. Set the source voltage at 2 volts and charge the capacitor C_1.

3. Move the switch so that the charged capacitor discharges through the circuit.

4. Observe the trace on the CRT. Can you detect any nonlinearity or the presence of a piecewise function?

5. Repeat the above steps for 4 V, 6 V, 8 V, 10 V, ..., 20 V.

6. Select a voltage which produces a CRT trace that gives a clear indication of oscillations occurring in both linear regions. What tells you that the trace originates from a piecewise function rather than a "soft" spring function? Explain the difference.

7. Measure the period and amplitude of the oscillations in both linear regions. Which region has the larger period? What is this period change telling you?

8. Notice that there is a small region where the trace gradually changes from one region to the other. See if you can locate one or two oscillations in the change-over region. What are the amplitudes and periods of these oscillations?

9. Using the given values of the capacitors and the bend voltage, construct a graph similar to that shown in Fig. 5.3. Use Maple file X05BSTFT.MWS to transform the graph to its "soft" spring form. What values does the Maple file give for the coefficients a and b? What ratio of C_1 to C_2 would produce a better fit between the soft spring and piecewise linear approximation?

10. Using the values for a and b just determined, calculate the values for α and β defined in Eq. (5.10).

11. After the values for α and β are known, use Maple file X05DUFF.MWS to solve Eq. (5.10) numerically and plot the solution.

12. Compare the experimental amplitudes and periods with those produced by the Maple file.

13. Use Maple file X05PIECE.MWS to solve Eq. (5.5) numerically with the piecewise expression (5.7) inserted. Plot $q(t)$ vs. t.

14. Repeat all of the above steps for a value of C_1 twice the value of C_2.

15. Repeat the above steps for a value of C_1 five times the value of C_2.

Things to Investigate:

- Use the circuit shown in Fig. 5.5 to produce a larger knee or bend voltage.

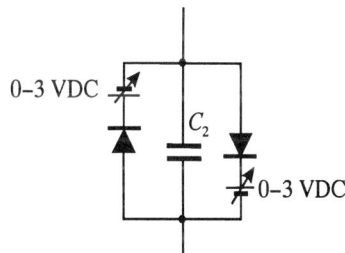

Figure 5.5: Batteries increase the bend voltage.

(This method of producing a nonlinearity is also used in Experimental Activity 25, Chua's Butterfly.) Does this larger bend voltage produce a more easily discernible nonlinearity?

- Repeat the experiment with three or more capacitors. The capacitance of each capacitor should have a different value. All but one of the capacitors should have diodes placed in parallel with them.

- Use three or more identical capacitors but construct different bends by using the set-up shown in Fig. 5.5. Each parallel diode circuit should use different potentials (battery values) to create more steps

Experimental Activity 6

Tunnel Diode Negative Resistance Curve

Comment: This experimental activity is easy with no surprises and should take about 1 hour to complete.

References:

1. *Nonlinear Physics with Maple for Scientists and Engineers, 2.3.2.*

2. [BN92] This is a useful book that covers the basic principle of electronics and solid state circuit devices.

Object: This investigation shows how the characteristic *I–V* curve of a tunnel diode can be determined.

Theory: Semiconductor devices, although they are usually made to operate in a linear region, are inherently nonlinear. This activity explores the performance of one of the strangest of all the nonlinear semiconductor devices, the tunnel diode. The reason that tunnel diodes are so strange is that their *I–V* curves have a negative resistance region. In order to help the electronics neophyte understand the construction and operation of diodes, a brief introduction to normal and tunnel diodes is presented below. If the reader is familiar with the theory and operation of ordinary and/or tunnel diodes, you may skip to the procedure.

A. Ordinary Diodes Semiconductors are made from elements, e.g., silicon or germanium, that have four electrons in their outermost (valence) shell. To make intrinsic (pure) semiconductors conduct better, small amounts of valence 3 or valence 5 elements are added to their crystal lattice. When these elements are appropriately added, it is known as doping the semiconductor. The rest of this theory explains how these two different doping elements can be used to produce two different types (p, n) of conduction (materials) from intrinsic semiconductors and how diodes are constructed from them.

Figure 6.1 represents an atomic model of a pure silicon (Si^{4+}) semiconductor. Silicon atoms prefer to have eight valence electrons rather than the normal

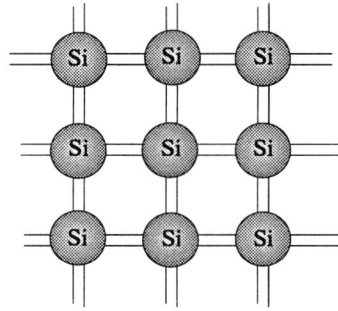

Figure 6.1: The straight lines joining the atoms represent shared electrons.

four in their outer shells, so each atom bonds covalently by sharing one of its electrons with its closest neighbor. In Fig. 6.1 each straight line represents a bond and a shared electron.

If the silicon has a small percentage of its crystalline atoms replaced (doped) by valence 3 atoms, such as boron (B^{3+}), empty bonding sites (holes) are created. When the semiconductor is doped to produce holes, it is known as a p-type semiconductor. Fig. 6.2 shows a schematic model of a p-type semiconductor. The holes can accept stray electrons that drift or wander in from the neighbor-

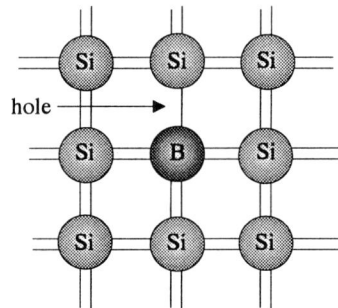

Figure 6.2: Boron, a hole increasing dopant, has been added to the silicon.

ing atoms. An electron leaving one bond to migrate to an empty hole makes it appear as if a hole is moving in the lattice in the opposite direction. The movement of a hole is quantum mechanically equivalent to the movement of a positive charge. As an everyday analogy of the movement of a hole, consider a line of cars stopped at a red light. Assume that between one of the cars and the next stopped car is a large gap. When the light changes to green and the cars (electrons) start to move forward, the gap (hole) appears to move backward. In p-type semiconductors it is more conventional to think of the holes moving rather than the electrons.

If a valence 5 atom such as phosphorus (P^{5+}) is used to dope the silicon atoms, free electrons are created when the bonds accept only four electrons per atom and reject one of the donor's five electrons. Fig. 6.3 shows an example of this kind of doping. These free electrons, like the positive holes in p-type

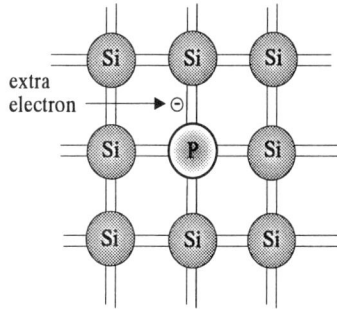

Figure 6.3: Phosphorus, an electron increasing dopant, added to the silicon.

material, are available to drift through the semiconductor. Semiconductors with free electrons are known as n-type semiconductors.

A diode is created when a piece of p-type material is joined to a piece of n-type material. When this joining is properly done, the free electrons in the n-type material drift across the p–n junction and fall into the holes that exist near the junction. This depletes the number of holes near the junction and reduces the number of free electrons available for further depletion. The electron migration causes the net charge of the p-type material to become negatively charged near the junction and at the same time causes the n-type material to become positively charged near the junction, thus producing an electric field $\vec{E}$ as shown in Fig. 6.4. Because the region on both sides of the junction is now

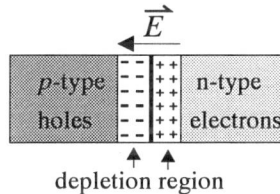

Figure 6.4: A junction p–n diode showing the depletion region.

depleted of its majority carriers, it is known as the depletion region. Examining Fig. 6.4 shows that the direction of this created electric field acts to hinder further movement of electrons or holes across the junction. As the depletion region and net charge in the region continue to grow, the strength of the electric field and the junction potential increases. The strength of the electric field increases until it is strong enough to stop the migration of electrons across the junction. For silicon, a junction potential of around 0.70 volts[1] (Ge, 0.35 volts) is produced when equilibrium is established.

The circuit symbols for a diode and a tunnel diode along with the shapes of the common enclosing cases are shown in Fig 6.5. The diode's arrow represents

---

[1]This 0.70 volts is the diode switch voltage mentioned in the last activity.

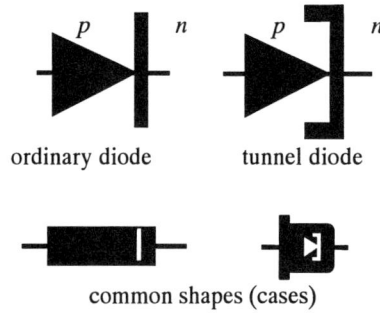

Figure 6.5: Diode and tunnel diode circuit symbols and common cases.

the p-type (the anode) material and the vertical bar represents the n-type material (the cathode). Another way of remembering this convention is that the conventional current flows more easily in the direction of the arrow. Cylindrical diodes have their cathode end marked by a line which encircles the cylinder.

If a normal diode is connected to an external voltage source with the polarities as shown in Fig. 6.6, the electric field across the diode's junction is increased. This acts to widen the diode's normal depletion region and makes it

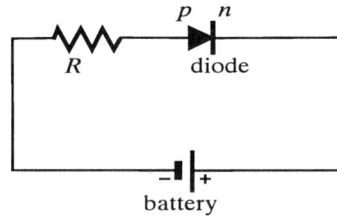

Figure 6.6: A p–n diode in reverse bias.

even harder for current to cross the junction. A diode wired in this way is said to be in reverse bias. In the reverse bias mode no (or very little) current flows in the circuit.

If a normal diode is connected to an external voltage source with the polarities as shown in Fig 6.7 it is said to be in the forward bias mode. In the

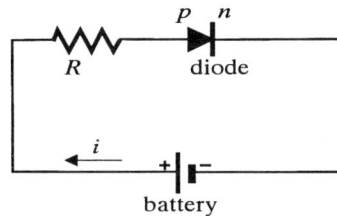

Figure 6.7: A p–n diode in forward bias.

forward bias the external voltage source, if large enough (> 0.7 V for Si, > 0.35 V for Ge), overcomes the internal junction potential. When this occurs, large numbers of electrons and holes can continuously move across the junction. Sustained electron and hole flow across the junction will produce a relatively large current in the external circuit.

For the circuit shown in Fig 6.7, if the battery is replaced by an alternating voltage source, the current will be rectified (becomes a pulsating DC current). The current is large when the AC polarity produces a forward bias, and there is no current when the AC polarity produces a reverse bias. This signal rectification is one of the most important uses of an ordinary diode.

B. Tunnel Diodes Tunnel diodes, although constructed similarly to ordinary diodes, function quite differently and are usually made of germanium instead of silicon. Leo Esaki[2] introduced the tunnel diode in 1958. Ordinary diodes have doping concentrations of about 10^{20} m^{-3}. Tunnel diodes have their doping concentrations increased 10,000 or more times to 10^{24} m^{-3}. The larger doping creates many more holes and free electrons in the respective n and p semiconductors. With the larger doping a much smaller depletion region is created. The space needed to hold the electric charge required to establish the junction potential is much smaller than that needed by a normal diode. The increased doping reduces the width of the depletion region from about 10^{-6} m to about 10^{-8} m. When the width of the depletion region is this small and when the correct forward bias is used to propel the electrons toward the energy barrier created by the depletion region, the electrons are able to tunnel quantum mechanically through, rather than jump over, the energy barrier. Because this quantum tunneling takes place near the speed of light, a tunnel diode has the capability of being used as a very fast switching device. However, this high speed switching can cause problems for circuit designers because at high frequencies the stray capacitances and inductances contained in the wires and contact points become more important and may introduce unwanted signals into the circuit.

The solid line shown in Figure 6.8 is a representative plot of the current (I_d) through a tunnel diode vs. the potential difference (V_d) across the diode.

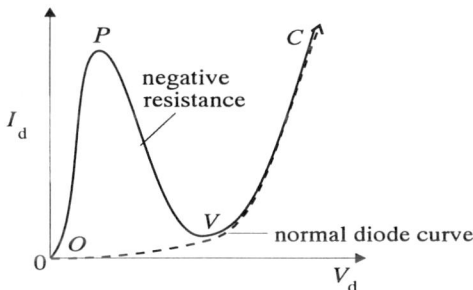

Figure 6.8: Current vs. potential for a tunnel diode.

---

[2]Esaki, Giaever, and Josephson shared the 1973 Physics Nobel prize for their work on quantum mechanical tunneling in semiconductors.

The line OPV represents the tunneling portion of the curve. The dashed line represents the current through an ordinary germanium diode as a function of the potential difference across the diode. The region of negative slope, P to V on the curve, is the region of negative resistance. Different tunnel diodes will have different operating characteristics but the shapes of the curves are similar.

The main purpose of this experimental activity is to generate a specific response curve for the tunnel diode under investigation. The result should be a plot similar to the one shown in Fig. 6.8. Successfully completing this experimental activity makes the next experimental activity dealing with tunnel diode oscillators easier to understand and perform.

Procedure:

1. Wire the circuit shown in Figure 6.9. The circuit uses the tunnel diodes[3] IN3718 or IN3719, but other tunnel diodes could be used.

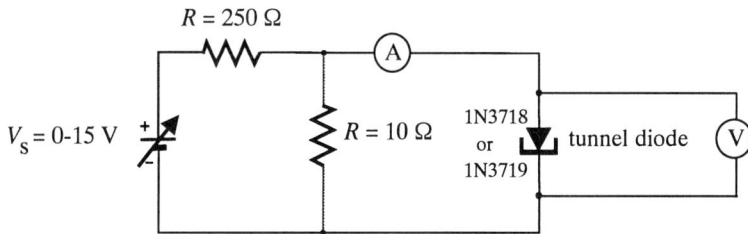

Figure 6.9: Circuit used to produce tunnel diode response curves.

2. Slowly increase the supply voltage (V_s), recording the potential drop across the diode and the current through the diode. Tunnel diodes are easily damaged so be careful not to exceed the manufacturer's recommendations for maximum current. If the diodes 1N3718 or 1N3719 are used the maximum current is 50 mA.

3. As the potential drop across the diode is slowly increased, don't be alarmed when a further increase in voltage produces a decrease in current. You have entered the region of negative resistance.

4. Keep increasing the source voltage until the diode is back in its positive resistance range. Continue to record the potential drop across the diode and the current through the diode. Do not exceed the maximum current of the diode.

5. Plot the values for V_d along the abscissa and the values for I_d on the ordinate. Your curve should look somewhat like Figure 6.8.

6. What is the average value of the negative resistance? The reciprocal of the slope is equal to the resistance.

---

[3]Tunnel diodes may be purchased from Germanium Power Devices Corporation, 300 Brickstone Sq., Andover, MA 01810 (508-475-1512).

7. What is the difference between your curve and the curve given in Fig. 6.8?

Things to Investigate:

- Replace the 10 Ω resistor with a 50 Ω resistor and rerun the experiment. Explain what happens when the peak voltage is exceeded. After the peak voltage is reached, slowly decrease the source voltage and watch for any sudden jumps in the measured values. What replacement resistance would stop this hysteresis?

- Assuming the equation of the tunnel diode curve is $I_D = A_1 V^3 + A_2 V^2 + A_3 V$, determine the values for the constants A_1, A_2 and A_3. If you have used the previous best-fit Maple files, you might wish to use one of these files to help with this task.

- Assuming the reference point is moved to the midpoint of the negative resistance portion of the curve, the equation can be approximated by $i = -av + bv^3$. What are the values for the constants a and b?

Experimental Activity 7

Tunnel Diode Self-Excited Oscillator

Comment: This is an easy experimental activity and should not take more than 2 hours to complete. Doing this experiment will help you understand many of the concepts in the text, e.g., limit cycles, self-excited oscillations, relaxation oscillations, etc.

References:

1. *Nonlinear Physics with Maple for Scientists and Engineers, 2.3.2.*

2. Experimental Activity 6: Tunnel Diode Negative Resistance Curve.

3. [BN92] An electronics text with an excellent section on tunnel diodes.

Object: To investigate self-excited oscillations governed by the Van der Pol (VdP) equation with special attention to the changes in shape of the oscillations as one of the control parameters is varied.

Theory: This theory is relatively short because the development of the Van der Pol (VdP) equation governing the oscillations produced by this experimental activity is found on the provided Maple file X07TUN.MWS. This file also permits a comparison of the experimental results with the theoretical predictions as various control parameters are changed.

This experimental activity uses the circuit given in Problem 2.18 of the accompanying text. The characteristic I–V curve for a tunnel diode is shown in Fig. 7.1. Experimental Activity 6 shows how to produce this curve. An approximate equation for the I–V curve in the neighborhood of the operating point I_s, V_s is given in the text as

$$i = -av + bv^3 \tag{7.1}$$

where $a = 0.050$ and $b = 1.0$ for the tunnel diode 1N3719. Using Kirchhoff's current and voltage rules produces the unnormalized VdP equation

$$\ddot{v} - (\alpha - \beta v^2)\dot{v} + \omega^2 v = 0 \tag{7.2}$$

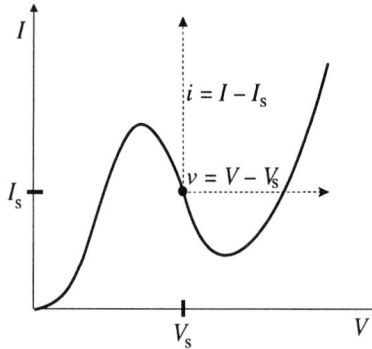

Figure 7.1: Tunnel diode I–V curve.

where $\alpha = \frac{1}{C}(a - \frac{1}{R})$, $\beta = \frac{3b}{C}$, and $\omega = \frac{1}{\sqrt{LC}}$. In this activity the resistance R is varied to see how this control parameter affects the shape of the oscillations.

Procedure:

1. Wire the circuit as shown in Figure 7.2. When using tunnel diodes, par-

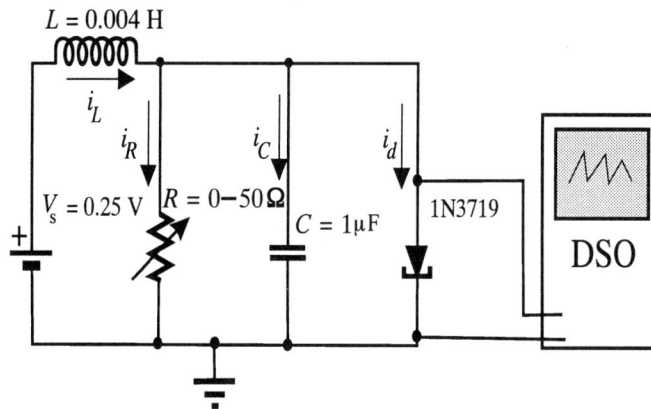

Figure 7.2: The circuit.

asitic capacitances and inductances can cause circuit problems, so it is advisable to keep the wire leads as short as possible. If a different tunnel diode is used the values for a and b will have to be calculated. The variable resistor should be a resistance box that allows you to use a number of different and accurate resistances.

IMPORTANT:

- The power supply must be able to deliver a constant voltage of $V_s = 0.25V$. Older power supplies may not be able to maintain a constant output voltage.

- The inductor must have a low resistance so that the circuit does not contain a resistance in the wrong location. The PSSC air core solenoid which has an inductance of 0.004 H and a resistance of less than 1.0 Ω works fine.

2. Turn on the circuit and adjust the power supply to the correct voltage. Set R to 50 Ω.

3. Adjust the digital storage oscilloscope (DSO) until a trace showing an oscillation appears on the CRT.

4. Calculate the theoretical oscillation frequency $\nu = \frac{1}{2\pi\sqrt{LC}}$.

5. Using the DSO, measure the oscillation frequency to ensure the circuit is producing the correct frequency.

6. Slowly decrease R to see what affect it has on the shape of the CRT trace. As R decreases, does the CRT trace better approximate a sine wave?

7. Make a sketch of the wave shapes for four widely separated R values.

8. What is the critical value for R that makes the trace disappear?

9. Starting with an R value somewhat larger than the critical value produce a steady-state CRT trace. Quickly lower the R value to a value a little lower than the critical value. How does the CRT trace respond?

10. Use Maple file X07TUN.MWS to compare the CRT trace with that predicted by Maple.

Things to Investigate:

- Why does the circuit continue to oscillate when the capacitor is removed from the circuit? What happens to the frequency? What does this tell you about the internal construction of the tunnel diode?

- Repeat the above experiment with different inductors.

- Examine Figure 6.4 in the accompanying text to see what a relaxation oscillation looks like. Does this experiment produce this shape, and if so for what values of R?

Experimental Activity 8

Forced Duffing Equation

Comment: This investigation should not take more than 2 hours to complete.

References:

1. *Nonlinear Physics with Maple for Scientists and Engineers, 2.7.1.*

2. [Bri87] This article contains a similar experiment.

3. [DFGJ91] This article contains a discussion of an inverted pendulum and how it can be used to model the Duffing equation.

4. [Pip87] This excellent book has a detailed analysis of the inverted pendulum and as a bonus contains a large number of ideas for additional experiments.

Object: To use an inverted pendulum to investigate the motion, especially the chaotic motion, of an object governed by the forced Duffing equation.

Theory: The importance of the Duffing equation in nonlinear physics cannot be overstated. For example, the accompanying text uses the Duffing equation in a variety of ways and for a variety of reasons:

1. The simplicity of the Duffing equation is deceptive in that the physical behavior it models is remarkably complex;

2. The Duffing equation is a generalization of the linear differential equation that describes damped and forced harmonic motion. This allows many analogies and comparisons to be drawn between the physical behavior modeled by these linear and nonlinear equations;

3. The Duffing equation appears in many disguises and aliases, e.g., forced, unforced, hard spring, soft spring, negative stiffener, and nonharmonic oscillator;

4. The Duffing equation can be used to demonstrate nonlinear behavior such as

- the oscillation period varies with the amplitude;
- it exhibits jump phenomena and hysteresis;
- it exhibits bifurcation phenomena displaying the period doubling route to chaos;
- it can be used to generate very simple but instructive Poincaré sections.

The negative stiffener Duffing equation is studied in this activity.

The negative stiffener is created by using the inverted pendulum as shown in Figure 8.1. The inverted pendulum is made from steel tape similar to that

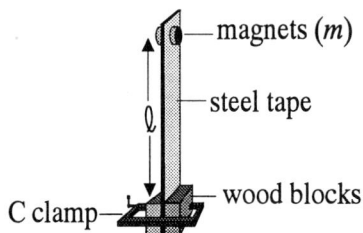

Figure 8.1: The inverted pendulum (Euler strut).

used to make airtrack bumpers or to wrap shipping cartons. Two or more strong (neodymium) magnets are used to provide the mass (m) at the top of the pendulum. The magnets permit the effective length (ℓ) of the pendulum to be easily changed. Using Fig. 8.2 as a reference, an approximate expression for

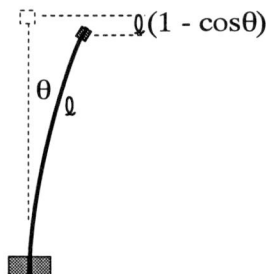

Figure 8.2: Inverted pendulum at angle θ.

the potential energy of the pendulum is

$$V = \frac{k\theta^2}{2} - mgy, \tag{8.1}$$

where $y = \ell(1 - \cos\theta)$ and k is the stiffness constant of the steel spring. The potential energy is assumed to be zero ($V = 0$) at the top of the trajectory. If, in Eq. (8.1), the $\cos\theta$ term is written as a power series

$$\cos\theta = 1 - \frac{\theta^2}{2!} + \frac{\theta^4}{4!} + \dots, \tag{8.2}$$

the potential energy is approximately equal to

$$V = \frac{1}{2}(k - mg\ell)\theta^2 + \frac{1}{24}(mg\ell)\theta^4. \tag{8.3}$$

The terms with exponents greater than four are ignored because it is assumed that θ is small. Notice that in Eq. (8.3) the coefficient of the θ^2 term can be positive or negative. A plot of the potential energy (V) vs. (θ) for a varying ℓ is shown in Fig. 8.3. The double well curves represent values for ℓ that give

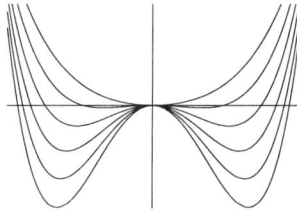

Figure 8.3: Potential (V) vs. angle (θ).

negative coefficients for the θ^2 term, i.e., for $mg\ell > k$. As ℓ increases, the depth of the double wells increases. The double well curves have an unstable saddle point at $\theta = 0$. Although this activity deals with motion under the control of a double well potential, if $mg\ell < k$, a single well potential characteristic of the hard spring Duffing equation is created.

Now that the approximate expression for the inverted pendulum's potential energy has been derived, the Lagrangian $L = T - V$ can be used to produce the differential equation that describes the motion. Neglecting damping and forcing,

$$L = \frac{m\ell^2\dot{\theta}^2}{2} - \frac{1}{2}(k - mg\ell)\theta^2 - \frac{mg\ell}{24}\theta^4, \tag{8.4}$$

and applying the Lagrange equation

$$\frac{d}{dt}\left(\frac{\partial L}{\partial \dot{\theta}}\right) - \frac{\partial L}{\partial \theta} = 0 \tag{8.5}$$

an approximate differential equation that describes the motion of an inverted pendulum is produced:

$$m\ell^2\ddot{\theta} + (k - mg\ell)\theta + \frac{1}{6}(mg\ell)\theta^3 = 0. \tag{8.6}$$

On adding a damping term $b\dot{\theta}$ and a forcing term $\tau_0 \cos(\omega_d t)$, where τ_0 is the maximum torque and ω_d is the driving frequency, Eq. (8.6) becomes

$$m\ell^2\ddot{\theta} + b\dot{\theta} + (k - mg\ell)\theta + \frac{1}{6}(mg\ell)\theta^3 = \tau_0 \cos(\omega_d t). \tag{8.7}$$

Expressing Eq. (8.7) in a more conventional form gives

$$\ddot{\theta} + 2\gamma\dot{\theta} + \left(\frac{k}{m\ell^2} - \frac{g}{\ell}\right)\theta + \frac{1}{6}\frac{g}{\ell}\theta^3 = \alpha_0\cos(\omega_d t). \qquad (8.8)$$

Eq. (8.8) is the equation used to describe the motion in this activity.

This Activity has two procedures. Procedure A is simpler and does not require as many measurements as Procedure B.

Procedure A: Simpler of the Two Procedures

1. Construct the apparatus shown in Fig. 8.4. If you do not have this equipment there is an alternate way of forcing the pendulum. Figure 8.5 shows how a Helmholtz coil can be used to power the pendulum. The inverted pendulum can be made from the steel tape used for airtrack bumpers (recommended), from a hacksaw blade, or from the steel tape used to wrap

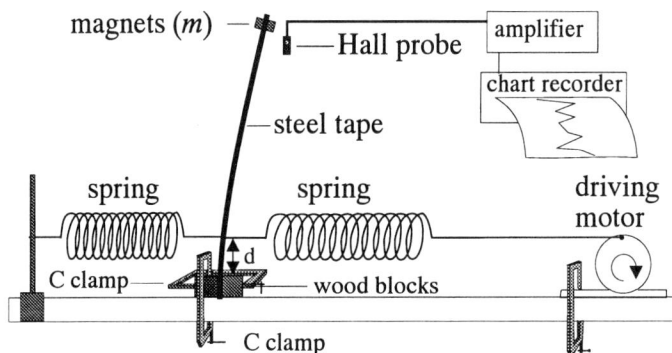

Figure 8.4: Apparatus for the inverted pendulum.

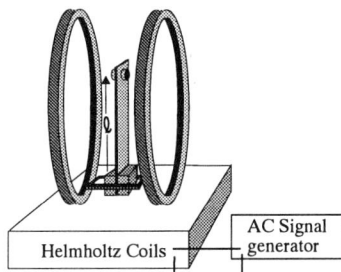

Figure 8.5: An alternate driving method.

shipping cartons. A length between 0.20 m and 0.40 m is satisfactory.

2. Connect the two coiled springs to the inverted pendulum. A file can be used to make small v grooves along the edges of the tape. These grooves

can be used to hold the springs at a fixed spot. Springs that have a spring constant of about 1.0 N/m are recommended. The 10 cm long coiled springs that come with the airtrack are appropriate.

3. When the motor is not running, adjust the position of the magnets until the steel tape is just into its two-well mode. The two-well mode can be recognized by having the magnets stay, when placed, in either one or the other of the two equilibrium positions found on each side of the vertical center position. The apparatus shows two wood blocks clamping the pendulum at its base. The steel tape can be moved up and down between these blocks if adjustments to its length are required. Later, when the motor is running, small adjustments in the springs' tension can be made by sliding the motor back and forth along the table as needed.

4. Place the Hall probe near the top of the inverted pendulum's trajectory. This can be done by taping the Hall probe to a wooden meter stick and then moving the meter stick to the desired position.

5. The output from the Hall probe should be connected to a storage oscilloscope or to a paper chart recorder.

6. Start the motor running. Adjust the frequency or driving amplitude of the motor. A very low frequency ($\nu \approx 0.1 \leftrightarrow 1.0$ Hz) is probably required. After the motor is running you might wish to adjust its horizontal position to make the motion of the pendulum behave more symmetrically. Try to produce an amplitude that remains small enough to keep the Hall probe measuring a signal strength over the whole of the oscillation.

7. The pleasure gained in watching this weird and wonderful motion is, in itself, a worthwhile procedure, so if pressed for time just observe the motion. Attempt to identify period doubling, quadrupling, etc., and confirm your observations by examining the chart recorder or CRT trace.

8. Make a number of different chart recordings. Vary the driving frequency and driving amplitude.

9. It is easy to adjust the motor's frequency or amplitude to produce chaotic motion.

Procedure B: A More Quantitative Procedure

1. Use the same apparatus as in Procedure A.

2. Measure the mass of the small powerful magnets which are attached to the top of the steel tape.

3. Before connecting the coiled springs to the inverted pendulum, measure their stiffness constants (k_s). Strong springs $k_s \approx 10$ N/m should be used if a hacksaw blade is used for the steel strip. Weaker springs can be used if steel bumper tape is used to make the inverted pendulum.

4. Connect the coiled springs to the inverted pendulum.

5. Locate the critical point of the inverted pendulum. Measure d (the distance from the pivot to where the coiled springs are attached to the pendulum) and ℓ.

6. Calculate the linear damping coefficient (γ). An approximate value for the damping coefficient γ can be found by determining the time $t_{1/2}$ it takes the amplitude to halve. Then use the equation $\gamma = \frac{\ln(2)}{t_{1/2}}$ to calculate γ.

7. Calculate the value for the stiffness constant of the inverted pendulum by finding the critical point where the pendulum is moving from a single well to a two-well potential. At this point $k = mg\ell$.

8. Calculate the maximum force $F = k_s x_0$ where x_0 is the maximum amplitude of the driving motor and k_s is the coiled spring constant.

9. Calculate the maximum torque, $\tau_0 = Fd$, where d is the distance from the pivot to the point where the springs are attached to the inverted pendulum; see Fig. 8.4.

10. Connect the Hall probe to the chart recorder or a digital storage oscilloscope.

11. With the inverted pendulum at rest in one of its equilibrium positions, start the motor. Try to produce small oscillations so that the Hall probe can detect the pendulum's magnetic field over the full range of its oscillation.

12. Use Maple file X08DUFF.MWS to compare the chart recorder's plot with the plot produced by the file.

Things to Investigate:

- Investigate the motion in the two different potentials (single well and double well), but without using the driving motor to force the motion. Can you detect any nonlinear behavior? What kind of behavior should you be looking for?

- Use different steel tapes to make the inverted pendulum.

- When the pendulum is in its single well mode (hard spring), does it exhibit hysteresis?

- Limit the amplitude of the oscillation by adding an additional nonlinear repelling potential. Fasten two small repelling magnets near the end points of the oscillations. Can you still detect nonlinear behavior?

- How is this experiment changed if the inverted pendulum is inverted? Is this equivalent to negative gravity?

Experimental Activity 9

Focal Point Instability

Comment: This investigation should not take more than 2 hours to complete.

References:

1. *Nonlinear Physics with Maple for Scientists and Engineers*, 3.2.

2. [For87] Page 308 gives the circuit used in this activity.

3. [Pip87] This book contains a similar circuit to the one used in this activity. This is an excellent book, easy to read, and it contains many thought-provoking ideas.

Object: To investigate a focal point instability produced by an electrical circuit.

Theory: A damped undriven harmonic oscillator can be described by the following linear second order homogeneous differential equation:

$$\ddot{q} + 2\gamma\dot{q} + \omega_0^2 q = 0, \tag{9.1}$$

where γ is the damping coefficient. For $0 < \gamma < \omega_0$, the solution is

$$q = Ae^{-(\gamma t)} \cos(\omega t + B), \tag{9.2}$$

where A and B are the integration constants determined by the initial conditions, and $\omega = \sqrt{\omega_0^2 - \gamma^2}$. A plot of Eq. (9.2) is shown in Fig. 9.1. This

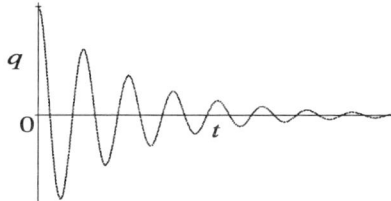

Figure 9.1: Positively damped ($\gamma > 0$) harmonic oscillator.

figure shows that the amplitude (q) of the oscillations exponentially decays as the time (t) increases. For a positive damping coefficient, energy is continuously dissipated by the system.

For the less frequently encountered and more unlikely case of having a negative damping coefficient ($\gamma < 0$), the solution to Eq. (9.1) for $|\gamma| < \omega_0$ is

$$q = Ce^{(|\gamma|t)}\cos(\omega t + D) \tag{9.3}$$

where C and D are integration constants. A plot of Eq. (9.3) is shown in Fig. 9.2. In this plot the amplitude (q) of the oscillations grows exponentially from

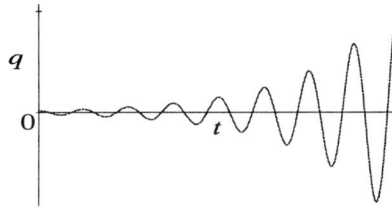

Figure 9.2: Negatively damped ($\gamma < 0$) harmonic oscillator.

some very small initial value as the time (t) increases. Energy is continuously absorbed by the system.

Phase plane plots for both of these solutions are shown in Fig. 9.3. For Eq. 9.2, the trajectory winds onto a stable focal point (F) as shown in plot A.

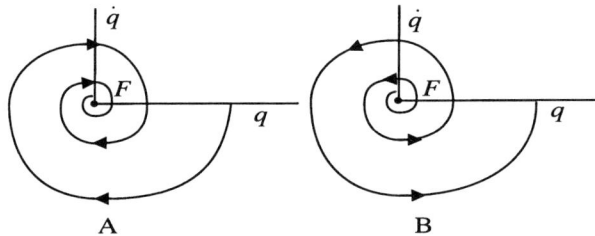

Figure 9.3: Phase plane trajectories for positive (A) and negative (B) damping coefficients.

Plot B is the phase plane plot for Eq. (9.3) and illustrates that the oscillations unwind from an unstable focal point. The negative damping coefficient means that any movement, however tiny, acts as a seed to start the oscillations. Once the oscillations have started their amplitude increases until the oscillator self-destructs or some other factor limits its growth. Real or actual examples which produce such a growth are rarely encountered in linear physics. Situations which give negative damping coefficients are hard to imagine. The presence of a negative damping coefficient indicates the oscillator absorbs energy from its surroundings and when encountered is very surprising, witness the collapse of the Tacoma Narrows Bridge. (Try to think of other physical examples of where γ is negative.) This activity provides an opportunity to investigate the phenomenon of a negative damping coefficient and its associated focal point instability.

In this activity an electric circuit is used to produce oscillations that behave as if a negative damping coefficient is present. The schematic of the circuit is shown in Fig. 9.4. The circuit contains a positive resistance (R_e), a negative

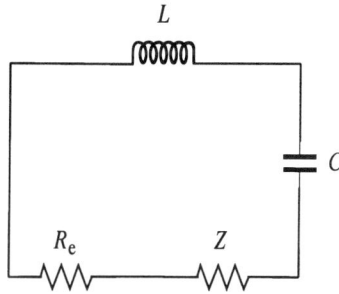

Figure 9.4: Circuit to produce an unstable focal point.

resistance ($Z = -r$), an inductor (L) and a capacitor (C). (Don't worry if you do not know how to purchase a negative resistance, all will be made clear.) Kirchhoff's voltage law says that the sum of the potential drops around the loop must sum to zero, so

$$V_L + V_{R_t} + V_C = 0, \tag{9.4}$$

where $V_L = L\frac{d^2q}{dt^2}$, $V_R = R_t\frac{dq}{dt}$, $V_C = \frac{q}{C}$, and $R_t = R_e + Z = R_e - r$ is the total resistance. Thus, the differential equation relating electrical charge (q) to time (t) is

$$\ddot{q} + \frac{(R_e - r)}{L}\dot{q} + \frac{1}{LC}q = 0. \tag{9.5}$$

Comparing Eq. (9.1) and Eq. (9.5) gives the relationship $2\gamma = \frac{(R_e - r)}{L}$ and $\omega_0^2 = \frac{1}{LC}$. Now if r is greater than the positive resistance (R_e) the circuit is negatively damped and the oscillations, once started, will grow exponentially. If the initial conditions for the circuit are $q = 0$ and $\dot{q} = 0$, the circuit is stable and quiescent. However, if the initial conditions have even the smallest nonzero values, the charge begins to oscillate and the amplitude of the oscillations exponentially increases. The term "unstable focal point" is well chosen. In this experiment, circuit noise will provide the tiny signal needed to seed, or initiate, the oscillations.

The negative resistance $Z = -r$ in this activity is produced by an operational amplifier (op-amp) wired as shown in Fig. 9.5. The triangular circuit symbol is the op-amp. To understand how an ideal op-amp functions, consider it to be a black box that operates under the control of a few simple rules:

1. The negative sign ($-$) on the inverting terminal of the op-amp does not indicate a polarity. It indicates that the output signal, if connected directly, is 180 degrees out of phase with the input signal;

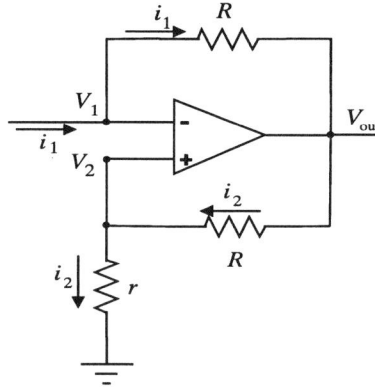

Figure 9.5: Op-amp circuit to produce negative resistance.

2. The positive sign (+) on the noninverting input of the op-amp does not indicate a polarity. It indicates that the output signal, if connected directly, is in phase with the input signal;

3. No current flows into the inverting (−) or noninverting (+) inputs. The ideal op-amp is considered to have an infinite input impedance;

4. The voltages at the inverting (−) and noninverting (+) inputs are equal. This is a consequence of the above rule;

5. An op-amp has no (or very low) output impedance.

The op-amp's power supplies are not shown in Fig. 9.5.

To show how the op-amp acts as a negative resistance, consider Figure 9.5 and the above enumerated rules. The input impedance (Z) of the circuit is defined by the equation

$$Z = \frac{V_1}{i_1}. \tag{9.6}$$

The value for Z is to be found in terms of r. Since the op-amp draws no current, the current i_1 does not enter the op-amp but must travel through the top resistor (R) so

$$V_1 - V_{\text{out}} = i_1 R. \tag{9.7}$$

Solving Eq. (9.7) for i_1 and substituting into Eq. (9.6) gives

$$Z = \frac{V_1 R}{V_1 - V_{\text{out}}}. \tag{9.8}$$

The output voltage is given by

$$V_{\text{out}} = i_2 r + i_2 R \tag{9.9}$$

and since $V_1 = V_2$ (Rule 4), $i_2 = \frac{V_1}{r}$ and then Eq. (9.9) becomes

$$V_1 - V_{\text{out}} = -\frac{V_1 R}{r}. \tag{9.10}$$

Substituting Eq. (9.10) into Eq. (9.8) gives

$$Z = -r, \tag{9.11}$$

which shows that an op-amp wired in this configuration acts as a negative resistance with a magnitude r. It should also be noted that if r is replaced with a capacitor (inductor), the negative sign changes the impedance from a capacitance to an inductance or vice versa.

When the sub-circuit shown in Fig. 9.5 is connected to a capacitor (C),

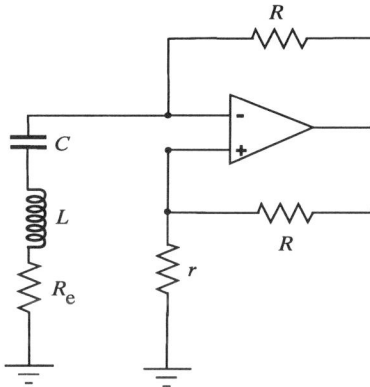

Figure 9.6: Equivalent circuit to Fig. 9.4.

inductor (L), and resistance (R_e) as shown in Fig. 9.6, the circuit is equivalent to the circuit shown in Fig. 9.4. This equivalent circuit permits the total resistance to be easily and slowly changed from negative to positive. The speed with which the resulting oscillations exponentially grow or decay can be controlled by changing the value (positive or negative) of the total resistance.

Procedure:

1. Wire the circuit shown in Fig. 9.7. Any dual inline 8-pin mini dip op-amp should work but μA741 is a nice choice. It is much easier if the circuit is mounted on a breadboard that comes with the +15 and −15 volt power supply. The values of the inductor and capacitor are not critical, but the solenoid should not have a ferromagnetic core. A digital storage oscilloscope makes it easier to study the trace.

2. Slowly adjust the variable resistor (r) until its value is approximately equal to the resistance of the external circuit. A dial resistance box, if graduated in small enough divisions, can be used for the variable resistor.

3. When the resistor (r) is larger than R_e the trace should explode onto the CRT. What is this critical value for r? The signal should reach a maximum value of around 28 volts.

4. If the resistance of r is just a wee bit larger than R_e, you will see a trace with a slow but exponential increase in the amplitude of the oscillation.

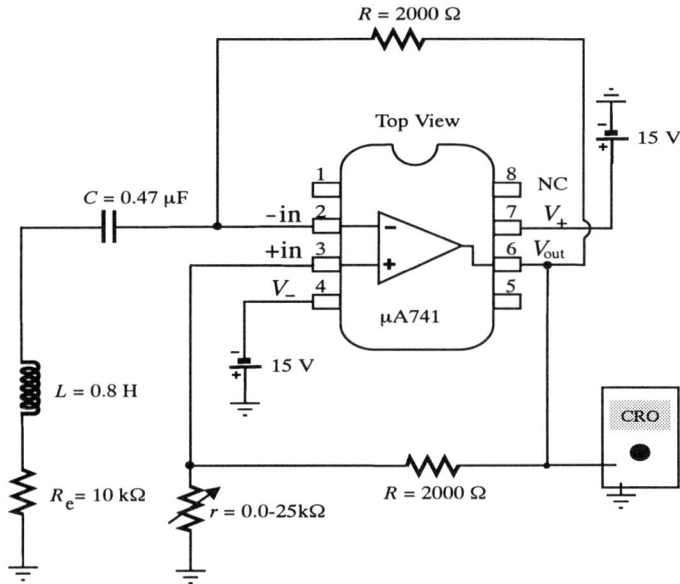

Figure 9.7: Circuit to produce unstable focal point.

With proper adjustment of r you should be able to make this growth take a relatively long time ($\approx$ 30 s or more) to reach its saturation voltage.

5. With the trace in its saturated state, reduce r by a small amount and watch the exponential decay of the trace.

Things to Investigate:

- Have some fun with this very unusual circuit. Adjust the values anyway you wish but keep the resistors marked R above 1500 Ω. This keeps the output current from exceeding the op-amp's specification of 25 mA.

- Try using an inductor with a ferromagnetic core. Can you detect any nonlinear effects when you do this?

- Connect a signal generator in series with the external circuit. Using a very low power signal and a negative resistance below the total external resistance look for a signal trace that shows the transient and steady state solutions of a forced oscillator. (Make sure you include the output impedance of the signal generator in your total external resistance.)

- Explain why the signal generator should not be used to input a signal when the negative resistance is larger than the external resistance of the circuit.

- If you have access to a SPICE program such as MICRO-CAP IV try simulating the operation of the circuit shown in Fig. 9.7 on the computer.

Experimental Activity 10

Period of a Compound Pendulum

Comment: This experimental activity should take less than 1 hour to complete.

References:

1. *Nonlinear Physics with Maple for Scientists and Engineers, 4.2.4.*

2. [Cro95] This article provides ideas for supporting an oscillating rod, diagrams illustrating different pendulum configurations and the equations for calculating their periods, and suggestions for additional and more complex investigations of rigid rod oscillations.

Object: To measure the period of a pendulum that has a large initial amplitude.

Theory: In this experiment, the period of a pendulum swinging with a large amplitude is measured and then compared with the theoretical value calculated by a provided Maple file X10PEN.MWS. To ensure that the pendulum swings in a circular arc, a compound pendulum (a meter stick) is substituted for the more traditional string and bob pendulum. Figure 10.1 is a sketch of the pendulum. This type of pendulum allows for the use of large initial angles, where a simple string and bob pendulum would fall rather than swing. If the air resistance and the sliding friction at the pivot are ignored, the differential equation for the torque on the pendulum (meter stick) is

$$I\frac{d^2\theta}{dt^2} = -rmg\sin\theta \tag{10.1}$$

where I represents the moment of inertia, m the mass of the pendulum, g the gravitational field strength, and r the distance from the pivot to the center of mass. The moment of inertia of the stick can be found by the parallel axis theorem to be

$$I = \frac{m\ell^2}{12} + mr^2, \tag{10.2}$$

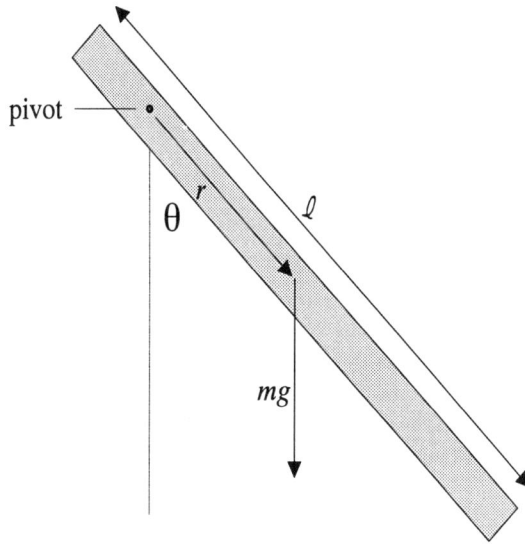

Figure 10.1: The compound pendulum.

where ℓ is the length of the stick. The above differential equation can be written in the standard simple pendulum form

$$\frac{d^2\theta}{dt^2} + \omega_0^2 \sin\theta = 0, \qquad (10.3)$$

with

$$\omega_0 = \sqrt{\frac{rg}{r^2 + \frac{\ell^2}{12}}} \ . \qquad (10.4)$$

The small angle frequency ω_0 should be calculated before doing the experiment. The period for small oscillations can be calculated using $T = \frac{2\pi}{\omega_0}$, while for large oscillations

$$T = \frac{4}{\omega_0} K(k). \qquad (10.5)$$

where $K(k)$ is the complete elliptical integral of the first kind, $k = \sin(\frac{\theta_{\max}}{2})$, and $\theta_{\max}$ is the maximum angle of oscillation.

Procedure:

1. Set up the apparatus as shown in Fig. 10.2. Suspend the meter stick so that the distance from its center of mass to the fulcrum is about 40 cm.

2. Calculate the theoretical period for small oscillations.

3. Measure the small oscillation period. Compare the measured and theoretical values.

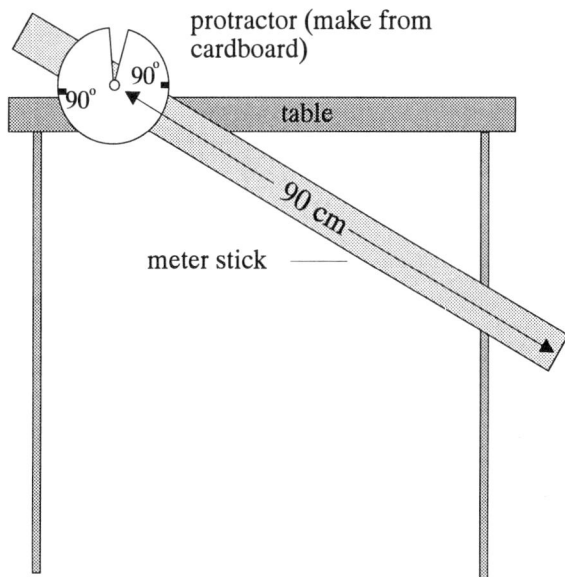

Figure 10.2: Apparatus for measuring the period of a compound pendulum.

4. Start the pendulum with a large initial amplitude of sixty degrees. With a stop watch, measure the period for one complete swing. Note the maximum return angle. If friction makes the return amplitude much smaller than the starting angle, use an average value for the amplitude. Make several measurements to reduce the error. Use Maple file X10PEN.MWS to compare the measured and calculated values for the period.

5. Repeat Step 4 using angles of 80°, 100°, 120°, 140° ....

Things to Investigate:

- At what minimum angle can measurement discern a difference between the small oscillation period and the actual period?

Experimental Activity 11

Stable Limit Cycle

Comment: After the circuit has been constructed, this investigation should not take more than 2 hours to complete.

References:

1. *Nonlinear Physics with Maple for Scientists and Engineers*, 6.1.

2. [HCL76] The Wien bridge equations may be found in this article.

3. [HN84] An article that discusses a similar experiment.

4. [Mal93] Contains an excellent explanation of the Wien bridge oscillator.

Object: To use a Wien bridge circuit to investigate a signal that has its origin in an unstable focal point and ends in a stable limit cycle. The signal will be studied in its transient and steady state regimes.

Theory: This activity uses a Wien bridge oscillator to produce a stable limit cycle. Elementary electronic courses study the Wien bridge circuit because it can produce very precise and stable sine waves. In these courses the nonlinear properties of the Wien bridge oscillator are rarely mentioned or explored. The oscillations produced by a Wien bridge are autonomous, i.e., the forcing function is not periodic or time dependent. The final periodic motions that result from autonomous systems are known as limit cycles. This activity should be of special interest to electrical engineers who might want to look at both the linear and nonlinear aspects of this important circuit.

The Wien bridge oscillator uses a counterbalancing negative and positive feedback to produce a very stable sine wave signal. When the circuit is first turned on, the positive feedback is larger than the nonlinear negative feedback. Any white noise that is present in the circuit provides the correct frequency seed to cause the circuit to leave its unstable focal point. As the amplitude of the oscillations increases, the current increases through a nonlinear resistance which usually is an incandescent bulb or other thermistor. As the current through the nonlinear resistance increases, the negative feedback controlled by the resistance of this resistor grows until it balances the positive feedback. When this balance is attained the circuit has reached its limit cycle. The Wien bridge circuit is usually operated near its threshold and in this case the voltage vs. time limit cycle is a near perfect sine wave. The phase plane limit cycle is a circle (ellipse).

A schematic of the Wien bridge oscillator is shown in Fig. 11.1. The trian-

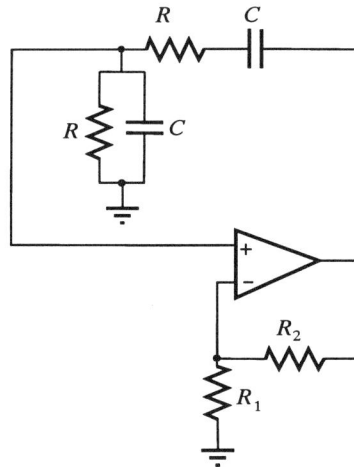

Figure 11.1: Wien bridge oscillator.

gular circuit symbol represents an operational amplifier (op-amp). In an effort to understand how the complete circuit in Fig. 11.1 functions, the circuit is first broken into its positive and negative feedback sections and then these sections are analyzed separately and in combination with each other.

The positive feedback section of the Wien bridge oscillator shown in Fig. 11.1 is reconstructed in Fig. 11.2. The equivalent circuit is also shown with the rectangular boxes representing the complex impedances Z_1 and Z_2. The positive

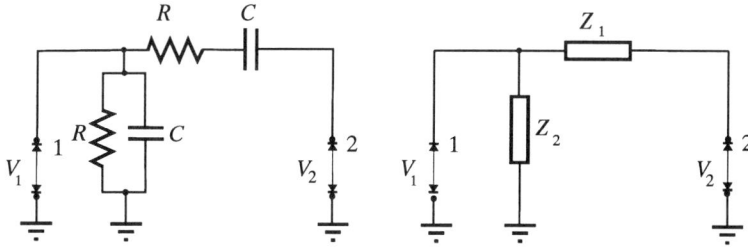

Figure 11.2: Positive feedback circuit and its equivalent.

feedback circuit is analyzed to see how much of the AC signal present at the op-amp's output (V_2) is sent back to the input (V_1). All signal strengths are measured relative to a ground level of 0.0 volts. An AC output signal at terminal 2 (V_2), causes a current (i) to flow through the complex impedances (Z_1, Z_2). The value for the current is

$$i = \frac{V_2}{Z_1 + Z_2},$$ (11.1)

where $Z_1 = R - jX$, $Z_2 = \frac{-jRX}{R - jX}$, and the capacitor's reactance is $X = \frac{1}{2\pi\nu C}$. The symbol j represents the imaginary number $\sqrt{-1}$ and ν represents the frequency of the AC signal. The strength of the AC signal between ground and terminal 1 is $V_1 = iZ_2$ so

$$V_1 = V_2 \frac{Z_2}{Z_1 + Z_2}.$$ (11.2)

In terms of R and X, Eq. (11.2) becomes

$$V_1 = V_2 \frac{(-jRX)}{(R^2 - X^2) - j(3RX)}.$$ (11.3)

Eq. (11.3) predicts that the maximum positive feedback occurs when $R = X$. Accordingly,

$$V_1 = \frac{V_2}{3}$$ (11.4)

which gives the maximum positive feedback of $\frac{1}{3}$ of the output voltage back to the input terminal.

Fig. 11.1 shows how this positive feedback subsection of the circuit is connected to the op-amp. If $R = X$, one-third of the output signal is fed back into the noninverting terminal (+) of the op-amp. A non-zero value for the positive feedback means that if an external signal, no matter how small, is present, the op-amp quickly goes to saturation. For example, assume the initial input signal had a unit value. The positive feedback would first increase it to $1 + 1/3$, then to $1.33 + 1.33/3$, then to $1.77 + 1.77/3$ ... and so on to saturation. Saturation occurs because the circuit cannot keep increasing the signal's amplitude (voltage) past the op-amp's capability.

The phase shift between the output and positive feedback input signals must also be known. If there was a phase shift between the input and output signal, the positive feedback could destructively interfere with the input signal.

Multiplying the numerator and denominator of Eq. (11.3) by the denominator's complex conjugate, the phase shift is given by

$$\tan\phi = -\frac{(R^2 - X^2)}{3RX}.$$ (11.5)

Eq. (11.5) shows that when $X = R$ there is no phase shift between V_2 and V_1, so the phase shift can be ignored. The positive feedback remains at 1/3 as previously calculated.

The frequency (ν) which makes $R = X = 1/(2\pi\nu C)$ is

$$\nu = \frac{1}{2\pi RC}.$$ (11.6)

All other frequencies produce phase shifts. The circuit's white noise will contain a frequency that corresponds to the resonance frequency. This frequency is quickly amplified until saturation is reached. Anyone who has heard a sound system accidentally amplify its own sound is aware of the effects of positive feedback. Something must be done to limit this electronic catastrophe. The Wien bridge oscillator uses a nonlinear negative feedback to control the linear positive feedback.

The negative feedback section of the Wien bridge circuit shown in Fig. 11.1 is shown in Fig. 11.3. Before attempting an explanation of how the negative

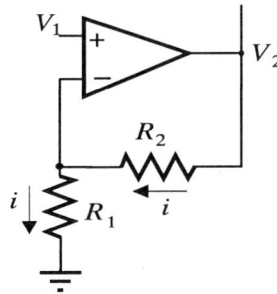

Figure 11.3: Negative feedback section of Wien bridge.

feedback section of the circuit operates, the rules that describe how an ideal op-amp function are reviewed:

1. The negative sign ($-$) on the inverting terminal of the op-amp does not indicate a polarity. It indicates that the output signal, if connected directly, is 180 degrees out of phase with the input signal;

2. The positive sign ($+$) on the noninverting input of the op-amp does not indicate a polarity. It indicates that the output signal, if connected directly, is in phase with the input signal;

3. No current flows into the inverting ($-$) or noninverting ($+$) inputs. The ideal op-amp is considered to have an infinite input impedance;

4. The voltages at the inverting ($-$) and noninverting ($+$) inputs are equal. This is a consequence of the above rule;

5. An op-amp has no (or very low) output impedance.

With these rules in mind, the amount of negative feedback is easily calculated. The voltage V_2 is

$$V_2 = i(R_2 + R_1).\qquad(11.7)$$

The voltage at the inverting ($-$) input of the op-amp is

$$V_- = iR_1 \qquad(11.8)$$

and since $V_- = V_+ = V_1$, the negative feedback (B) is the ratio of the output voltage to the input voltage and is equal to

$$B = \frac{V_2}{V_1} = \frac{(R_1 + R_2)}{R_1}.\qquad(11.9)$$

Normally, op-amps are considered to have a very large gain (100,000 or more) for low frequencies. The negative feedback given by B reduces the total gain to some smaller and more easily managed value.

Now, the total gain G is the product of the constant positive feedback of $1/3$ and the variable negative feedback (B), so

$$G = \frac{B}{3}.\qquad(11.10)$$

If in Eq. (11.9), $R_2 = 2R_1$ then $B = 3$ and the full circuit has a net gain of one. If $R_1 < 0.5R_2$, then $B > 3$ and the net gain is larger than one. If $R_1 > 0.5R_2$ then $B < 3$ and any signal present at V_2 decays to zero. Therefore the critical value for the negative feedback gain is three. Wien bridge circuits produce stable sine waves by designing the negative feedback portion of the circuit to maintain a value of 3 automatically. When $B=3$, $G=1$, and stability has been reached.

The Wien bridge is a nonlinear circuit because the R_1 resistor has a nonlinear I–V curve. This resistance is usually provided by an incandescent light bulb. The resistance of this bulb grows as the potential drop across the bulb increases as shown in Fig. 11.4. For small V, the resistance of the bulb is a "well behaved function" which can be Taylor expanded to give

$$R_1(V) = R_0 + R'(0)V + \frac{1}{2!}R''(0)V^2 + ...\qquad(11.11)$$

Since the slope is zero at $V = 0$, then for sufficiently small V the higher order terms can be ignored and

$$R_d \equiv R_1 - R_0 = kV^2,\qquad(11.12)$$

where $k \equiv \frac{1}{2!}R''(0)$ and R_d represents the dynamic or incremental resistance of the bulb. The symbol R_0 represents the resistance of the bulb when no current is flowing through it.

The rate of change of the dynamic resistance (R_d) is controlled by two competing mechanisms:

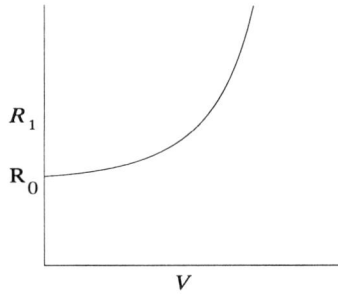

Figure 11.4: Resistance R_1 as a function of V.

- the forcing function due to the voltage drop V across the bulb and

- the tendency of the bulb to cool back to room temperature.

Assuming that the rate of change of R_d with time is proportional to R_d gives

$$\frac{dR_d}{dt} = -\frac{R_d}{\tau} + kV^2. \tag{11.13}$$

The first term on the right-hand side of the equal sign is the cooling contribution (note the minus sign) with the relaxation time τ a measure of the rate of cooling. The relaxation time is the time it takes the resistance to decrease from some value to $1/e$ of this value. The second term is the forcing contribution which follows from Eq. 11.12.

For the Wien bridge circuit to "turn on", the value for the initial resistance of the bulb ($R_1 = R_0$) must have a value less than $0.5R_2$ because Eq. (11.9), shows that in this case $B > 3$, which results in a net gain ($G = B/3$) that is greater than one. As the current through the bulb increases, its resistance (R_1) increases until the value for B decreases to 3. At this point the circuit is then producing a signal of constant amplitude and frequency—a near perfect sine wave. If the current would for some strange reason increase even further, the resistance (R_1) of the bulb would increase and the value for B would decrease below 3 and the gain would drop below 1 and the circuit would correct itself by lowering the value of the output signal until the value for B once again reaches three. The relaxation time (τ) of the resistor (bulb) is usually much larger than the natural period of the oscillator. If the relaxation time was smaller than the oscillation period, the corrections would distort the signal. Relaxation times for incandescent bulbs are usually in the 0.2–2.0 second range so the Wien bridge oscillator should have periods at least ten times smaller than these relaxation times. This means that the Wien bridge can produce reliable sine waves with a frequency of 5 Hz and up. This activity is the first of two that investigates the nonlinear properties of a Wien bridge circuit.

Procedure A: Producing the limit cycle.
Do not dismantle this circuit, it can be used in the next activity.

1. With a digital meter measure the initial resistance (R_0) of the incandescent bulb. A Radio Shack 12 V, 25 mA mini-bulb works very well for R_1. This

value for R_0 gives a rough idea of the value needed for R_2 to make the Wien bridge leave its focal point.

2. Wire the circuit shown in Fig. 11.5. Any generic op-amp from the 741 family should work, but μA741 is a reasonable and cheap op-amp to use. It is much easier if the circuit is wired on a breadboard that comes with the $+15$ and -15 volt power supplies. The values for the two resistances (R) or the two capacitors (C) are not critical, but make sure the capacitors (C) and resistors (R) have equal values and produce a frequency above 5 Hz. For example, $R = 1000\ \Omega$ and $C = 0.010\ \mu$F are reasonable values to use.

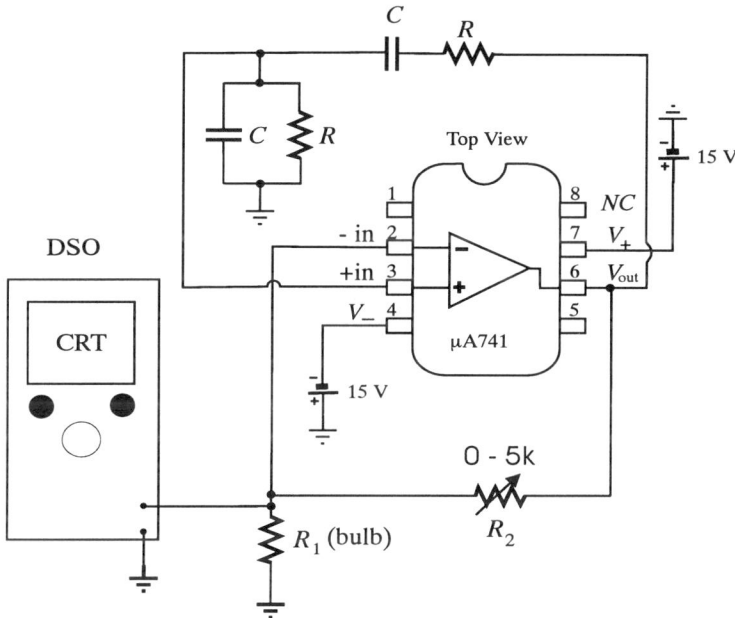

Figure 11.5: Wien bridge circuit.

3. For the resistor marked R_2 use a variable resistance box which permits the resistances to be read directly off the box. A storage oscilloscope makes it easier to study the trace.

4. Use Eq. (11.6) to calculate the value for the resonant frequency of your circuit.

5. Turn on your circuit.

6. Adjust R_2 ($R_2 \approx 2R_0$) until a sine wave is detected on the CRT. When the wave occurs you know that $2R_1 = R_2$.

7. What is the critical value for R_2? Calculate the value for R_1. What is the dynamic resistance R_d, where ($R_d \equiv R_1 - R_0$)?

8. Measure the value of the frequency of the limit cycle and compare it with the theoretical value.

9. After the CRT is displaying a sinusoidal trace, turn off the power to the circuit.

10. Turn the circuit back on and watch the signal reach its limit cycle.

11. Examine and try to explain the transient wave pattern.

12. Repeat the above steps for larger values of R_2.

If you wish, you may stop here or if you are still interested you might wish to try the following procedures.

Procedure B: Finding the relaxation time τ

1. Turn the circuit back on and observe the transient waves. Measure the modulation period of the transient wave and then calculate the modulation frequency.

2. Measure the relaxation time (τ). This is approximately the time for the modulated wave's amplitude to decrease to $1/e$ or $\approx 1/3$ of its initial value. These measurements can only be made in the transient regime.

Procedure C: Establishing the nonlinearity of R_1

1. Remove the bulb from the circuit.

2. Place the bulb in series with an ammeter and a DC power source.

3. Connect a voltmeter across the bulb.

4. Start with a potential difference of 0.05 volts and record the value of the current through the bulb. Calculate the resistance R_d.

5. Repeat the above step in increments of 0.05 volts up to a maximum of 1.0 volts.

6. Make a plot of R_d vs. V.

7. Assuming that Eq. (11.12) holds for lower voltages, what is the value for k? Does this relationship hold for larger voltages?

Things to Investigate:

- Have some fun with this circuit. Change the values for R and C and/or use a different type of incandescent bulb.

- Investigate the behavior of the circuit far from its threshold. This can be done by making $R_2 >> 2R_0$. Watch for different signal shapes and amplitude changes.

- Place the incandescent bulb in a beaker of cold water. How does this effect the transient portion of the wave?

Experimental Activity 12

Van der Pol Limit Cycle

Comment: This investigation should not be attempted unless Experimental Activity 11 has been completed. However, this activity is easy, useful and should not take more than 2 hours to complete.

References:

1. *Nonlinear Physics with Maple for Scientists and Engineers, 6.1.*

2. [DW83] This article contains numerical values that can be used with Maple to produce similar phase plane plots. It contains a nonlinear circuit that was once proposed to control the movement of an elevator, not really a good idea, but the circuit might provide an interesting idea for a self-designed experiment.

3. [HN84] Discusses a similar experiment using the Wien bridge oscillator governed by the VdP equation.

4. [HCL76] The source of the Wien bridge equation used in [HN84].

Object: To produce and investigate the limit cycle oscillations predicted by a Van der Pol-like equation.

Theory: This activity uses the same Wien bridge oscillator that was used in Experimental Activity 11. For convenience the circuit is shown again in Figure 12.1 As was seen in the last activity the bridge produced, when operated near threshold, self-excited oscillations that left an unstable focal point and grew into near perfect sine waves. The oscillatory signal was sampled by measuring the RMS voltage (V) across the resistor R_1. It is reasonable to assume the oscillations are governed by a differential equation similar to

$$\ddot{V} + 2\gamma\dot{V} + \omega^2 V = f(t), \tag{12.1}$$

where γ is the damping coefficient and $f(t)$ represents the seed noise to start the circuit running. Remembering that the dimensionless quality factor (Q), for

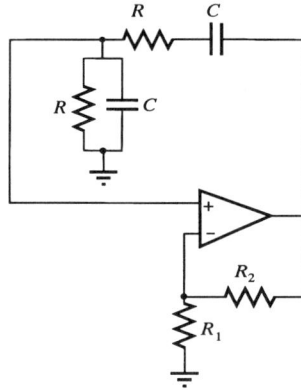

Figure 12.1: Wien bridge oscillator.

small damping, is given by $Q = \frac{\omega}{2\gamma}$ and for convenience letting $\Gamma = \frac{1}{Q}$ puts the above equation into the following form:

$$\ddot{V} + \Gamma\omega\dot{V} + \omega^2 V = f(t). \tag{12.2}$$

For the Wien bridge circuit to reach its limit cycle the coefficient Γ must be able to oscillate between a positive and a negative value. To understand how this might be accomplished, recall that in the last experiment the total gain of the Wien bridge circuit was shown to be

$$G = \frac{B}{3} \tag{12.3}$$

where the negative feedback B is

$$B = \frac{R_1 + R_2}{R_1}. \tag{12.4}$$

Further, recall that the value for B depends on the nonlinear resistor (the incandescent bulb) and that when the circuit is turned on the value for B starts with some value larger than 3 and therefore $G > 1$. When $G > 1$ the oscillations must grow so the coefficient Γ must be negative. The value for B decreases as the current through the resistor R_1 grows. As B decreases, G decreases below the value of 1. When $G < 1$ the oscillations decay so the coefficient Γ must be positive. One way of producing this sign oscillation in Γ is to write it as

$$\Gamma = -\beta(1 - \frac{1}{G}) = -\beta(1 - \frac{3}{B}) \tag{12.5}$$

where β is some positive constant.

Combining Eq. (12.2)and Eq. (12.5) produces

$$\ddot{V} - \beta\omega(1 - \frac{3}{B})\dot{V} + \omega^2 V = 0. \tag{12.6}$$

where the forcing term has been dropped ($f(t) = 0$) because this forcing term is only needed to start the oscillations, not to maintain them. Eq. (12.6) can be shown to be equivalent to the Wien bridge Eq. (2)

$$\ddot{V} + 9\omega(\alpha - \alpha_c)\dot{V} + \omega^2 V = 0. \tag{12.7}$$

found in [HN84]. To show the equivalence between Eq. (12.6) and Eq. (12.7), we note that $\alpha = \frac{1}{B}$, $\alpha_c = \frac{1}{3}$, and $\beta = 3$.

Substituting Eq. (12.4) into Eq. (12.6) and noting that the resistance R_1 (the incandescent bulb) is given by (from Experimental Activity 11 with $\frac{dR}{dt} \approx 0$ near equilibrium)

$$R_1 = R_0 + \tau k V^2, \tag{12.8}$$

we obtain

$$\ddot{V} - 3\omega\left(1 - \frac{3(R_0 + \tau k V^2)}{R_0 + R_2 + \tau k V^2}\right)\dot{V} + \omega^2 V = 0, \tag{12.9}$$

or

$$\ddot{V} - 3\omega\left(\frac{(R_2 - 2R_0 - 2\tau k V^2)}{(R_0 + R_2 + \tau k V^2)}\right)\dot{V} + \omega^2 V = 0. \tag{12.10}$$

Assuming

$$R_0 + R_2 \gg R_2 - 2R_0$$

and further that

$$R_0 + R_2 \gg \tau k V^2,$$

then Eq. (12.10) yields

$$\ddot{V} - 3\omega\left(\frac{(R_2 - 2R_0 - 2\tau k V^2)}{R_0 + R_2}\right)\dot{V} + \omega^2 V = 0. \tag{12.11}$$

Eq. (12.11) is an unnormalized Van der Pol equation. The purpose of this activity is to confirm experimentally the limit cycle oscillations predicted by the VdP-like equation. The values for k and τ measured in Experimental Activity 11 are required.

Procedure:

1. This is the same circuit as that used in Experimental Activity 11 so you should not have to rewire it. Make sure the same incandescent bulb is used. This saves you the effort of remeasuring the values for R_0, τ, and k.

2. Use $\nu = \frac{1}{2\pi RC}$ to calculate the value for the oscillation frequency of your circuit.

3. Turn on your circuit.

4. Adjust R_2 ($R_2 \approx 2R_0$) until a sinusoidal signal is just detected on the CRT. What is the critical value for R_2? What is the value for R_1? Why is the value for R_1 different than the value for R_0?

5. Adjust R_2 just past the critical value and make sure that a sine wave appears on the CRT. Measure the frequency and compare it with the theoretical value.

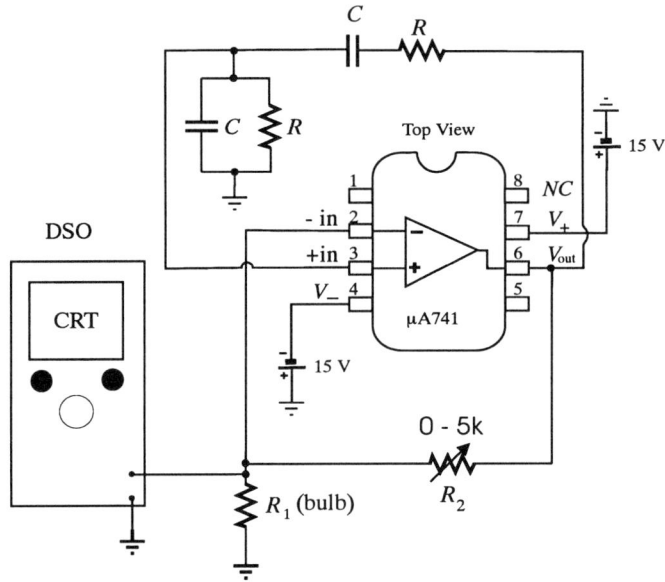

Figure 12.2: Wien bridge oscillator circuit.

6. After the trace has reached its limit cycle, measure or find the values for R_2 and V_{RMS}.

7. Increase the value of R_2 and repeat the above steps.

8. Place your experimental values into the Maple file X12VDP.MWS and check to see if the file reproduces the experimental results.

Things to Investigate:

- Will the threshold approximation fail for large values of R_2? What shape of trace would tell you the threshold approximations are becoming less valid?

- Change the values for R and C to alter the frequency ν. Repeat the above procedure and see if the near-threshold approximations are still valid for the same value of R_2.

Experimental Activity 13

Relaxation Oscillations: Neon Bulb

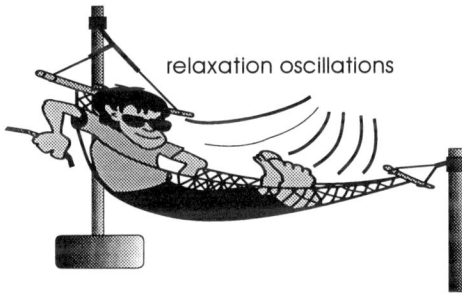

relaxation oscillations

Comment: This investigation should not take more than 1 hour to complete.

Reference:

1. *Nonlinear Physics with Maple for Scientists and Engineers, 6.2.*

Object: To investigate the relaxation oscillations produced by a constant voltage source applied to a circuit that contains a neon glow lamp wired in parallel with a capacitor.

Theory: Stable periodic oscillations that are caused by autonomous (time-independent) forcing functions are known as limit cycles. If the dependent variable exhibits fast changes near certain time values, with relatively slowly varying regions between, the oscillator is said to exhibit relaxation oscillations. In this experimental activity the negative resistance characteristic of a neon glow bulb in conjunction with a capacitor and constant voltage power supply is used to produce relaxation oscillations.

Figure 13.1 is a sketch of a neon glow lamp. The small bulb is about 1.0 cm long and contains two vertical metal electrodes which are separated by approximately 2 mm. The electrodes are surrounded by a moderately low pressure neon gas and when a potential difference is applied across the electrodes an electric

Figure 13.1: Neon glow lamp.

field is created. This electric field accelerates the electrons and neon ions inside the bulb. If the electric field is large enough, the electrons reach a speed that is sufficient to start an avalanche of electrons. The avalanche is caused by an electron knocking out of the neon atom a second electron, thereby providing two electrons for the next collision, and four for the next collision, and so on until saturation is reached.

Accordingly, a neon glow lamp is considered to have a very high resistance until the correct potential difference exists across the electrodes. At the critical potential difference—the firing voltage (V_f)—the presence of a seed electron creates an avalanche and the bulb begins to conduct. This conduction reveals itself by making the lamp glow. Once the neon bulb is glowing, its resistance decreases, so a current-limiting resistance is usually placed in series with the bulb. The current-limiting resistance is normally selected to have a value that permits the bulb to stay glowing without burning it out. This happens because as the current through the bulb increases and the potential drop across the bulb decreases, the potential drop across the resistance increases. If the correct resistance is selected, the bulb stays lit. If the current-limiting resistance is very large, the bulb turns off. If the current-limiting resistance is too small, the current can continue to increase until the bulb burns out. Because neon bulbs consume very little power and have a very long life they are often used to indicate if a device such as a power supply is turned on.

The neon bulb in this activity is used a little differently than that given in the description above. The circuit shown in Fig. 13.2 produces relaxation

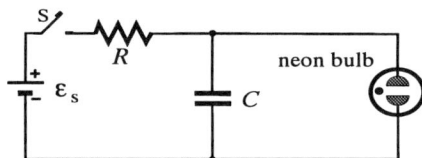

Figure 13.2: Neon tube circuit.

oscillations. When the switch (S) is closed, the capacitor slowly (relatively) charges through the large resistance (R). The potential drop across the capacitor increases until the bulb's firing voltage (V_f) is reached and the bulb turns on. As soon as the bulb turns on, its resistance drops and the capacitor rapidly discharges through the bulb. As the capacitor discharges, the potential drop across the capacitor (bulb) decreases until the extinction voltage is reached and the bulb turns off. When the bulb turns off, its resistance rapidly regains its large initial value (infinity), the capacitor begins to recharge and the limit cycle is in operation.

Fig. 13.3 shows the same cycle as a function of the voltage and the current. Be careful not to assume that the lengths of the line segments (DA, AB, BC, CD) are proportional to the time. As the voltage across the bulb increases

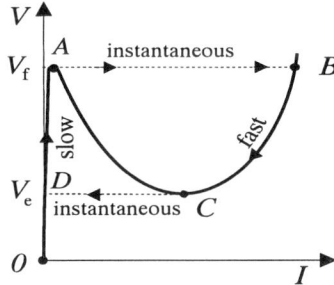

Figure 13.3: Voltage (V) vs. current (I) for a neon tube.

from 0 to A, no (or very little) current flows through the bulb. Once the firing voltage (V_f) is reached, the bulb's resistance drops dramatically and the bulb begins to conduct. This reduction in bulb's resistance allows for a rapid increase in the current through the bulb. This is represented by the movement from point A to point B. Now the circuit relaxes as the capacitor discharges and the current decreases from B to C. This discharge happens very rapidly because the resistance of the bulb when it is discharging is much smaller than the external resistance R. When the voltage across the bulb reaches the extinction voltage (V_e) at C, the bulb turns off and the current decreases from C to D. The voltage now restarts its slow climb from D to A. The cycle is complete. (This process is dependent on having a value for the external resistance R that is large enough to let the voltage drop across the bulb decrease past the extinction value. If the value for R was smaller than some critical value, it might be possible for the bulb to stay on because the extra current could be supplied by the battery.)

A voltage vs. time graph across the bulb is shown in Fig 13.4. The external resistance (R) is chosen to make the time to charge the capacitor through R much larger than the time to discharge the capacitor through the bulb. Eq. (13.1) gives the relationship between the voltage across the charging capacitor as a function of time

$$V = \varepsilon_s \left(1 - e^{-t/(RC)}\right), \tag{13.1}$$

and can be used to calculate the period (T) of the relaxation oscillations. In Eq. (13.1), ε_s represents source voltage which would be the maximum voltage across the capacitor if the bulb was removed from the circuit. The period $T = t_2 - t_1$ is calculated by solving Eq. (13.1) for the time t_1 by first rearranging

$$e^{-t_1/(RC)} = \frac{(\varepsilon_s - V_e)}{\varepsilon_s} \tag{13.2}$$

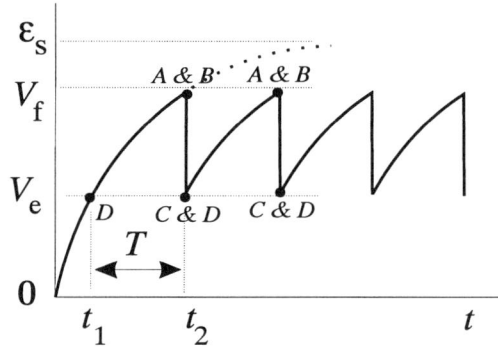

Figure 13.4: Voltage vs. time response curves for a neon tube.

and then solving for t_1 by taking the logarithm of each side of Eq. (13.2) to give

$$t_1 = -RC \ln \left| \frac{\varepsilon_s - V_e}{\varepsilon_s} \right|. \tag{13.3}$$

Similarly,

$$t_2 = -RC \ln \left| \frac{\varepsilon_s - V_f}{\varepsilon_s} \right|. \tag{13.4}$$

The period of the oscillation is given by $T = t_2 - t_1$, so

$$T = RC \ln \left| \frac{\varepsilon_s - V_e}{\varepsilon_s - V_f} \right|. \tag{13.5}$$

The main purpose of this activity is to see the shape of the relaxation oscillations shown in Fig. 13.4 and determine the period T.

Procedure A: Determining V_e and V_f

This procedure is for determining the firing voltage (V_f) and the extinction voltage (V_e). It can also be used to determine the voltage vs. current response curve for a neon bulb.

This activity uses high voltages, so please use caution. High voltage can be lethal. Before making changes to the circuit always turn the circuit off and wait a second or two for the capacitors to discharge.

1. Build the circuit shown in Fig. 13.5. Use a neon glow lamp of the class Ne-2. These bulbs have a variety of extinction and firing voltages but the firing voltage is usually between 70 and 110 volts and the extinction voltage between 50 and 70 volts. The variable DC power source (ε_s) should be able to produce a maximum of 200 volts. The 50k resistor should be at least 1 watt. You do not need to use the ammeter if you are only determining the extinction and firing potentials.

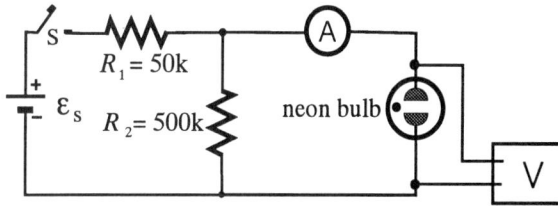

Figure 13.5: Apparatus to determine V_e and V_f.

2. Slowly increase the value of the source voltage and record the potential drop (V_f) across the neon bulb at the instant the bulb first begins to glow.

3. After the bulb lights, slowly decrease the source voltage and record the potential drop (V_e) at which the bulb turns off.

4. Repeat the above steps a couple of times to ensure that your values are reasonably accurate.

Making a plot of the voltage vs. current through a neon bulb.

1. Increase the voltage of the power supply until it makes the bulb turn on. Record the firing voltage.

2. Keep increasing the power supply's emf while at the same time record the current through the bulb and voltage across the bulb. Do not let the power supply's emf increase past 200 volts or the 50k resistor might burn out.

3. Record the current through and potential drop across the bulb as you slowly decrease the power supply's emf. Record the extinction voltage.

4. Make a plot of the voltage vs. the current.

Procedure B: Relaxation Oscillations

1. Build the circuit shown in Fig. 13.6. Use values for R of 0.50 $M\Omega$ and C of 0.47 μF (or a variable capacitor) and a variable DC emf (0–200 volts) power supply (ε_s).

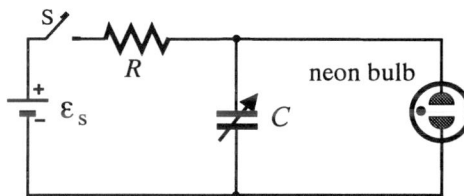

Figure 13.6: Relaxation oscillation circuit.

2. Turn on the power supply and slowly increase the voltage until the light begins to flash. Record the power supply's emf.

3. With a stop watch, measure the period of the flashing neon bulb. You might wish to adjust the value of C or R if the period is too short to measure easily with the stop watch.

4. Use Eq. (13.5) and the known values for V_f, V_e, and ε_s to calculate the period. How do the measured and calculated periods compare?

Procedure C: Observing the Phase Plane of the Relaxation Oscillator

1. Build the phase plane circuit shown in Fig. 13.7.

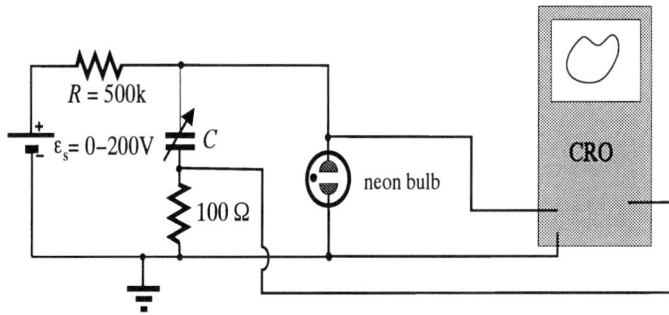

Figure 13.7: Phase plane circuit.

2. Adjust the source emf (voltage) until the bulb is flashing quite rapidly, i.e., $T \approx 0.20$ s.

3. Use the x–y and AC settings on the dual trace oscilloscope and observe the phase plane portrait.

Experimental Activity 14

Relaxation Oscillations: Drinking Bird

Comment: This investigation should not take more than 1 hour to complete.

Reference:

1. *Nonlinear Physics with Maple for Scientists and Engineers, 6.2.*

Object: To investigate the relaxation oscillations produced by the toy known as the drinking bird.

Theory: This activity uses an inexpensive toy, sold under the name Drinking Bird[1], as a relaxation oscillator. Figure 14.1 is a sketch of this well known toy. The toy normally operates by first wetting the cloth wrapped around the

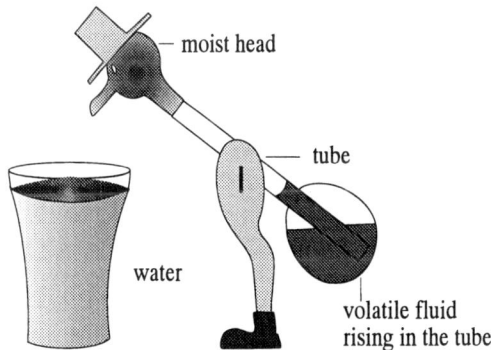

Figure 14.1: The drinking bird.

bird's head with water. The energy needed to evaporate the water, lowers the temperature of the head, and subsequently the air-pressure inside the head. The reduced pressure causes the volatile fluid contained within the tail of the bird

---

[1]Arbor Scientific, P.O. Box 2750, Ann Arbor, MI 48106-2750, phone 1-800-367-6695.

to rise through a vertical tube to the head. When the fluid reaches the head, its weight causes the bird to "drink" from the glass of water. The drinking action as shown in Fig. 14.2 performs three functions:

1. it keeps the bird's head wet by dipping it into the glass of water;

2. it lifts the lower end of the tube out of the liquid in the tail, thus equalizing the pressure inside the bird;

3. it provides a way for the fluid to flow back to the tail.

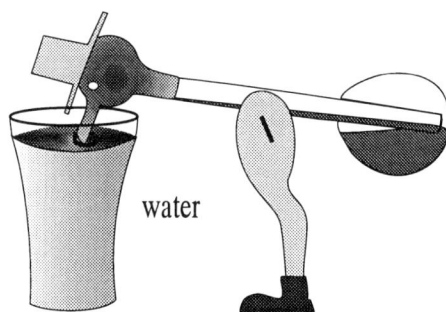

Figure 14.2: The drinking bird.

Accordingly, when the tube opens, the pressure equalizes, and the force of gravity pulls the fluid back to the tail and the bird upright. After the bird reaches its upright position, the fluid covers the bottom of the tube, the water that has soaked into the cloth wrapped around the bird's head evaporates, a pressure difference between the head and the tail is reestablished, and the bird is forced to drink once again. The relaxation oscillations have started. The bird's bobbing (drinking) is a relaxation oscillation because the energy source is not pulsating and because the bird's motion is characterized by short and long durations for different segments of its oscillation.

   In this activity, the bird's oscillations are produced in a slightly different way. The glass bulb which is the bird's tail is painted black (the colloidal solution used to make electrostatic pith balls works well as a paint) and a bright light is focused on the black tail. The heat produced by the absorbed light causes the fluid to rise to the head. The rising fluid forces the bird into its drinking position which pulls the tail out of the light. Once the bird is in the drinking position, the fluid flows back to the tail, the bird becomes erect, the tail reenters the light beam, and the oscillations have started.

   The strength of the light flux shining on the bird's glass bulb tail can be adjusted to alter the bird's bobbing rate. This is an easy to perform activity because the bobbing rate of the bird is slow enough to permit hand measurements of its period of oscillation. This activity qualitatively investigates the relationship between the strength of the light flux and how it controls the bird's vibrations.

Procedure:

1. Set up the following apparatus. The light source should be a 250 watt

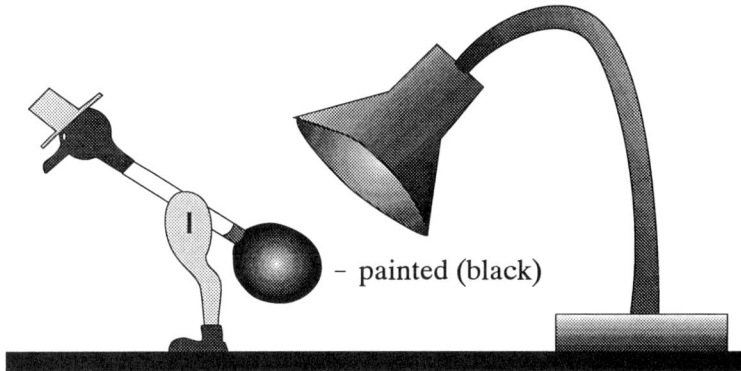

- painted (black)

Figure 14.3: Experimental setup.

nonfrosted light bulb. **Make sure the lamp can safely handle this size of bulb (wattage).** The light from an overhead projector also works well. A Fresnel lens may be used to focus the light on the bird's tail or to create a collimated beam of light.

2. Focus the light on the bird's tail so the bird begins to oscillate.

3. Measure the period of the bird's oscillations. Is the time between oscillations constant or does it vary with some pattern?

4. Repeat the above steps with different light intensities to modify the bird's oscillation rate.

5. Is there a critical flux (or distance from the light source) which makes the bird's oscillations begin or stop?

Things to Investigate:

- Plot a graph of the time and position of the bird's bobbing neck. Can you detect period doubling? Perform a power spectrum check on the graph. Use Experimental Activity 22 and its accompanying Maple file to help perform this procedure.

- Investigate how the combination of the light shining on the bird's tail and the bird's head bobbing into the water affects the rate of the oscillations.

- Use a dripping faucet to let water drip slowly onto the bird's head. Control the rate of the drips and investigate the effects on the bird's bobbing. Is there a critical drip rate?

Experimental Activity 15

Relaxation Oscillations: Tunnel Diode

Comment: This is an easy experimental activity and should not take more than 2 hours to complete.

References:

1. *Nonlinear Physics with Maple for Scientists and Engineers, 6.2.*

2. Experimental Activity 6: Tunnel Diode Negative Resistance Curve

3. [BN92] A useful book for extra information on tunnel diodes.

Object: To investigate self-excited relaxation oscillations produced by an electric circuit that consists of a constant energy source connected to a tunnel diode, inductor and resistor. To produce a tunnel diode's negative resistance curve from the relaxation oscillations.

Theory: Self-excited or auto-catalytic relaxation oscillators are highly nonlinear systems that spontaneously oscillate even though they are driven by a constant energy source. Self-excited relaxation oscillators have the ability to continuously absorb energy, but usually give off their energy in bursts or at specific locations in their phase space. The study of self-exited oscillators has played an important role in deepening the understanding of nonlinear systems. Relaxation oscillators, e.g., the Van der Pol oscillator, have been discussed in some detail in Reference 1. Self-excited oscillations are intimately connected with the formation of limit cycles.

The occurrence of self-excited oscillators in nature are many, one of the most important being your beating heart, and one of the most famous the wind driven torsional oscillations of the Tacoma Narrows bridge. The collapse of this bridge in 1940 was not due to the wind gusting in resonance with the natural frequency of the structure. It was due to the bridge's ability to absorb the wind's energy continuously and then release the energy in resonance with the bridge's natural frequency. Local residents observing the bridge being built, and after its completion, reported oscillations for winds blowing with speeds as low as 5 m/s (10 mph). The bridge was the destination of many Sunday drives, not just to see the marvelous structure, but also to feel its strange vibrations. Rumor has it that many of the local inhabitants predicted the collapse of this the world's largest unintentional swinging bridge long before it happened.

The electric circuit shown in Fig. 15.1 is used to explain how self-excited relaxation oscillations might occur. The circuit is designed so that it produces its operating point (V_0) near the middle of the unstable negative resistance region

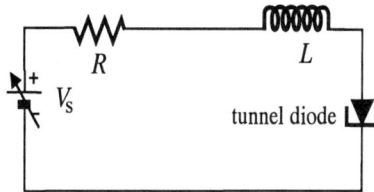

Figure 15.1: Relaxation oscillation circuit.

of the tunnel diode; see Fig. 15.2. When the voltage of the source shown in

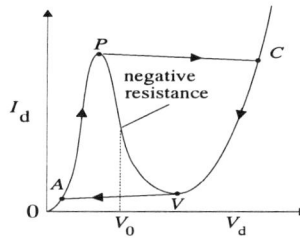

Figure 15.2: Current vs. potential for a tunnel diode.

Fig. 15.1 is made to increase very slowly toward V_0, the potential drop across the inductor can be ignored. The current continues to increase until the current in the tunnel diode reaches point P. If the voltage of the source is increased still further in an attempt to push the current even higher, the voltage drop across the diode must jump from P to C. The current cannot enter the negative resistance region because in that region the current would be decreasing. The jump from point P to C is very fast. During this jump the inductor quickly changes its polarity in an effort to maintain Kirchhoff's voltage rule. At point C the potential drop across the diode is larger than the source potential so the reversed potential of the inductor keeps the algebraic sum of the potentials around the circuit equal to zero. The reversed polarity will cause the current to decrease or relax toward V, hence the term *relaxation oscillations*. As the current reaches V the potential is still above the operating point (V_0) of the diode, so it must decrease even further. It cannot enter the negative resistance region because that would mean the current would increase, so the potential must quickly jump from V to A. During the jump the inductor quickly reverses its polarity to keep Kirchhoff's voltage rules intact. The potential of the source and inductor now cause the current to climb to P, and the process is repeated. This qualitative description presupposes that the inductance is sufficiently large to allow these large jumps from P to C and V to A. For small L it is possible for the system to oscillate in the negative resistance region. In this case the oscillations can appear very different. For very small oscillations in the negative resistance region they are sinusoidal. A voltage time plot for this relaxation oscillation process is shown in Fig. 15.3. The shape of the plot is characteristic

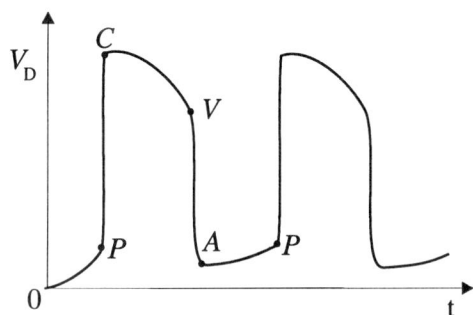

Figure 15.3: Voltage vs. time for the relaxation oscillator.

of many, but not all, relaxation oscillators. This activity should reproduce a replica of this plot.

** Tunnel diodes are very difficult circuit devices to get to function as predicted. The reason is that they like to oscillate at very high frequencies which make any small parasitic inductances and capacitances change the circuit's parameters. This is very bothersome to circuit designers. **

Procedure:

1. Wire either the circuit shown in Fig. 15.4 or Fig. 15.5. If the second circuit is used, the power supply must be able to maintain a set operating potential. Any tunnel diode should work, but if in doubt 1N3718 or 1N3719[1] is recommended.

2. If the circuit in Fig. 15.4 is used, adjust the variable resistance (R_1) to find an operating point in the negative resistance region of the tunnel diode's I–V response curve. You will know that you are in this region when the characteristic relaxation oscillation curve (Fig. 15.3) appears on the CRT.

3. If the circuit in Fig 15.5 is used, adjust the voltage between 0.05 volts and 0.30 volts to put the operating point somewhere in the negative resistance region.

4. Using the oscilloscope, measure the period of the oscillations.

---

[1] Tunnel diodes may be purchased from Germanium Power Devices Corporation, 300 Brickstone Sq., Andover, MA 01810 (508-475-1512).

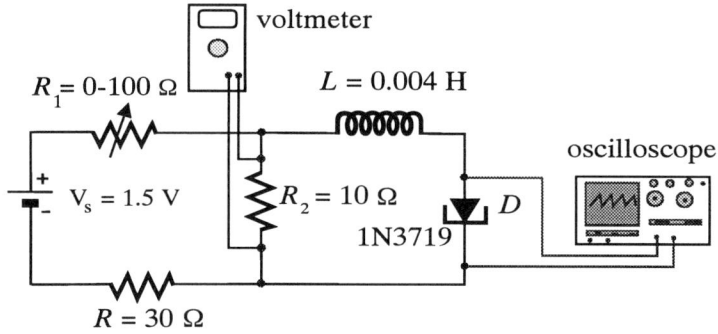

Figure 15.4: A tunnel diode circuit for producing relaxation oscillations.

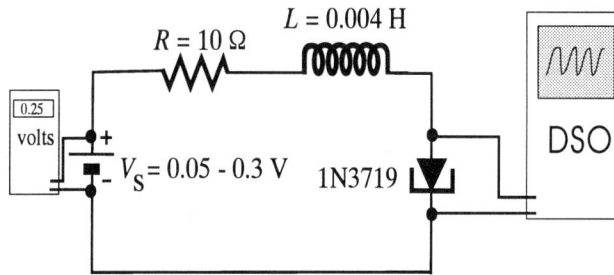

Figure 15.5: An alternate circuit for producing relaxation oscillations.

5. Measure the values for the voltages at points A, P, V, and C. If you completed Experimental Activity 6, compare these values with those obtained in that activity. If you did not complete Experimental Activity 6, sketch a rough plot of the response curve using the values measured here.

Things to Investigate:

- Remove the inductor from the circuit. Do relaxation oscillations still occur?

- Place a variable capacitor in parallel with the tunnel diode. Investigate how the relaxation oscillation is affected by changing the capacitance.

Experimental Activity 16

Hard Spring

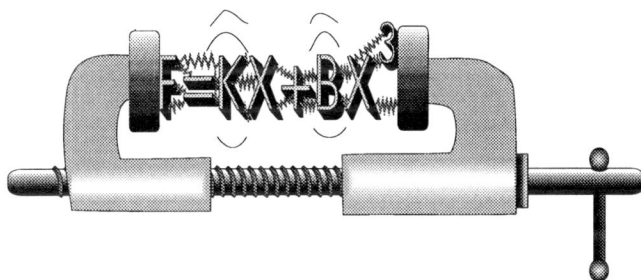

Comment: This experimental activity should take less than 1 hour to complete.

Reference:

1. *Nonlinear Physics with Maple for Scientists and Engineers, 7.1.*

Object: To find the mathematical relationship between a nonlinear spring's extension and the force required to produce that extension. To investigate the period vs. amplitude relationship for a hard spring.

Theory: The study of the motion that occurs under the control of a hard spring has played an important role in increasing the understanding of nonlinear physics. A spring is classified as a nonlinear hard spring when the restoring force increases faster than that given by Hooke's law. A graph of the force (F) as a function of the extension (x) for a typical hard spring is shown in Fig. 16.1. A possible equation for this curve is

$$F = ax + bx^3, \tag{16.1}$$

where a and b are positive constants. Damped oscillations governed by this nonlinear forcing function can be described using Newton's second law,

$$\ddot{x} + 2\gamma\dot{x} + \alpha x + \beta x^3 = 0. \tag{16.2}$$

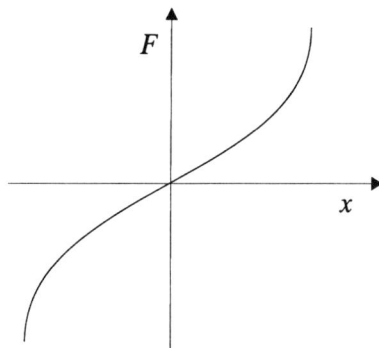

Figure 16.1: Force (F) and extension (x) for a hard spring.

where γ is the damping coefficient, $\alpha = \frac{a}{m}$, $\beta = \frac{b}{m}$, and m is the mass. A hard spring oscillator has a period that becomes smaller as the amplitude increases.

In this activity, a nonlinear hard spring arrangement is made by forming two ellipses from identical lengths of steel tape. (This type of steel tape is used to wrap cartons or comes with laboratory airtrack and is used to construct glider bumpers.) These bands of steel tape are used because they are highly nonlinear for large extensions. The reason for this nonlinearity is easy to deduce. For small extensions, extensions much smaller than the initial radius of the circular band of steel tape, the extension varies roughly as the pulling force. However, as the extension increases, the shape of the band of steel tape becomes more and more elliptical, and very large forces are now required to produce small increases in the extension. The limiting case would be where the circular spring is so deformed that it acts as two parallel steel tapes. Figure 16.2 shows how the two springs will be used in this experiment. Two springs are needed to

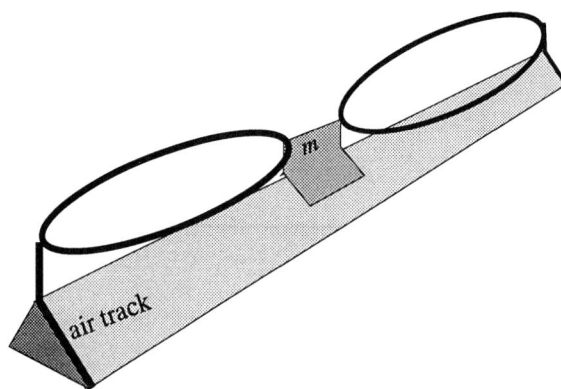

Figure 16.2: The nonlinear springs attached to an airtrack glider.

ensure that symmetrical restoring forces occur for both positive and negative extensions.

Procedure:

1. Connect the glider to the springs as shown in Fig. 16.3. The circular springs can be made from the same steel tape as used to make the airtrack bumpers. Masking tape is a convenient method of attaching the steel springs to the glider. Connect the springs so that they produce an oval shape.

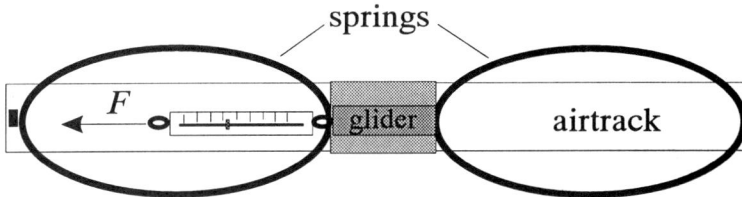

Figure 16.3: Top view: airtrack, airtrack glider and two nonlinear springs.

2. Attach a force meter to the glider. Measure and record the extensions for different forces. Make at least ten measurements on both sides (+ and −) of the equilibrium point.

3. Draw a graph of the force (F) as a function of the extension (x).

4. Find the values for the constants a and b in Eq. (16.1). You should use Maple file X16FIT.MWS to help find the best-fit values for the equation's constants. Just follow the instructions in the file.

5. Check the accuracy of the equation by measuring the period for various initial amplitudes. Use Maple file X16HRD.MWS to calculate the periods to see if there is agreement between the measured and theoretical values. (Measuring the period can be difficult because the springs are so heavily damped and therefore the oscillations decay very quickly. Measure the period for one or two oscillations only and then repeat a number of times to find the average.)

Things to Investigate:

- Repeat the activity using an airtrack glider of different mass.

- Repeat the activity using circular springs with different radii.

- Determine the value for γ by finding the time it takes for the amplitude of small oscillations to decrease by a factor of two.

Experimental Activity 17

Nonlinear Resonance Curve: Mechanical

Comment: This experimental activity should take less than 1 hour to complete.

Reference:

1. *Nonlinear Physics with Maple for Scientists and Engineers, 7.2.*

Object: To produce a nonlinear resonance response curve for a mechanical oscillator.

Theory: Nonlinear and linear resonance response curves are constructed by plotting the modulus of the steady state amplitude $|A|$ versus the driving frequency (ω). Figure 17.1 shows typical linear and nonlinear resonance response

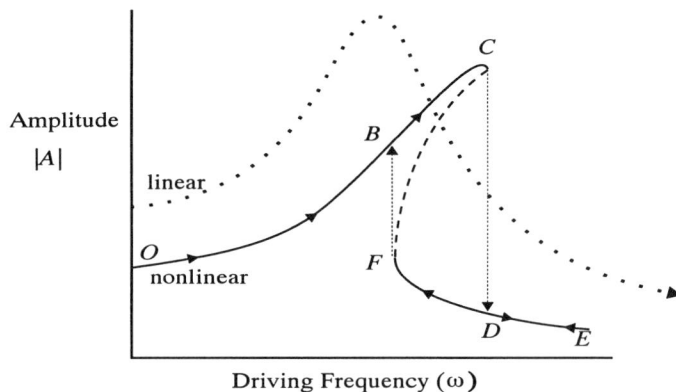

Figure 17.1: Linear and nonlinear resonance response curves.

curves. In the nonlinear case, as the driving frequency is slowly increased from O, a point C is reached where the value for the amplitude undergoes a sudden drop from C to D and then continues to point E and beyond. However, if the frequency is slowly decreased from point E, the amplitude will undergo a quick jump from F to B and then will continue along the curve to O. The frequency at which the jump from F to B occurs is different than that of the drop frequency. The phenomenon of following different system paths when moving in different directions is known as hysteresis.

The nonlinear hard spring is modeled by the equation that relates the pulling force (F) to the spring's extension (x) in the following manner,

$$F = ax + bx^3 \qquad\qquad (17.1)$$

where a and b are positive constants. A graph of the force as a function of extension is shown in Fig. 17.2. In this experiment the nonlinear hard springs

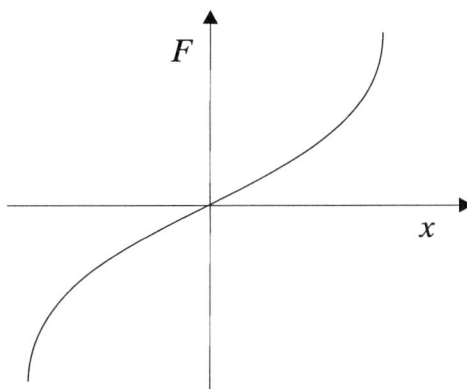

Figure 17.2: Force (F) and extension (x) for a hard spring.

are made by forming circles from lengths of steel tape. (This type of steel tape is used to wrap or secure cartons or a similar kind of tape comes with laboratory airtrack. It is used to construct the airtrack bumpers.) These circular bands are for large extensions highly nonlinear. The reason for this nonlinearity is easy to deduce. For small extensions, extensions much smaller than the radius of the circular band, the extension varies roughly as the pulling force. However, as the extension increases, the shape of the band of steel tape becomes more and more elliptical, and larger and larger forces are required to produce the same increase in the extension. The limiting case would be where the circular spring is so deformed that it acts as two parallel steel tapes. Figure 17.3 shows how the two springs will be used in this experiment. Two springs are needed to ensure that

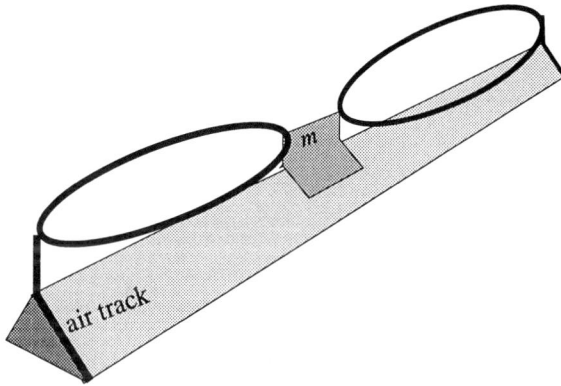

Figure 17.3: The nonlinear springs connected to an airtrack glider.

symmetrical restoring forces occur for both positive and negative extensions. A strong sinusoidal driving motor is connected to one of the springs. The motor is used to force the vibrating airtrack glider into its steady state. The motion of the vibrating glider can now be modeled by the forced Duffing equation

$$\ddot{x} + 2\gamma\dot{x} + \alpha x + \beta x^3 = A_0 \sin(\omega t). \tag{17.2}$$

The constants α, β, γ, and A_0 can be determined experimentally but are not needed in this activity.

Procedure:

1. Set up the following apparatus. The circular springs should be connected so as to produce an oval shape.

2. The driving motor should be connected to a variable voltage power supply so that the frequency of the motor can be adjusted.

3. With the motor running at a low frequency, measure the motor's frequency (or voltage) and the steady state amplitude of the air track glider. The circular springs are quite heavily damped (internally), so you should not have to wait too long for the glider to reach its steady state amplitude.

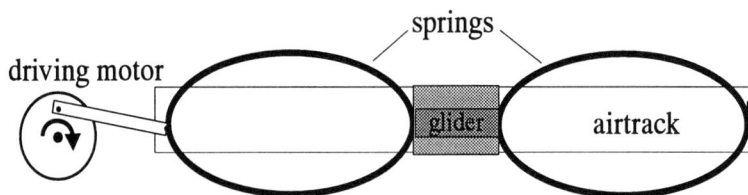

Figure 17.4: Two nonlinear springs connected to a forcing motor.

4. Increase the voltage (motor's frequency) a little and repeat the above measurement.

5. Repeat the above step until a sudden drop in the glider's amplitude occurs.

6. With the motor running at a frequency well above the drop frequency, measure the glider's amplitude.

7. Slowly reduce the motor's frequency, and for each reduction measure the glider's steady state amplitude.

8. Repeat the above step until a sudden jump in the glider's amplitude occurs.

9. Draw an amplitude–frequency graph for your data.

Things to Investigate:

- Repeat the activity with small strong magnets attached to the glider. (These magnets produce even stronger damping due to inducing a current in the airtrack. Lenz's law then can be used to explain the magnetic damping.) Does the hysteresis loop still occur at the same frequencies?

- Repeat the activity using the same circular springs but with a different initial elliptical (oval) setting. What is the effect on the data?

- Investigate the behavior of the resonance frequency of the apparatus.

- Look for examples of period doubling.

Experimental Activity 18

Nonlinear Resonance Curve: Electrical

Comment: This investigation should not take more than 2 hours to complete.

References:

1. *Nonlinear Physics with Maple for Scientists and Engineers*, 7.2.

2. Experimental Activity 5 explains how the piecewise linear circuit used in this activity functions.

3. [FJB85] This article shows how similar electrical circuits can be used to model the hysteresis effects of "hard spring" and "soft spring" oscillators.

Object: To investigate how the resonance frequency changes as a function of the forcing amplitude (voltage) for a piecewise linear (nonlinear) circuit. The nonlinearity is similar to that of a "soft spring" oscillator.

Theory: Resonance has played an important role in the studying and classifying of linear and nonlinear systems. Accordingly, one must be careful to understand how the resonance frequency differs for linear and nonlinear oscillations. For example,

1. a 1-dimensional linear system has one resonance frequency while a non-linear system can have an infinite number;

2. a linear system has one resonance frequency regardless of the pumping force, while nonlinear systems do not.

This activity explores and investigates these two properties.

A modification of the electric circuit in Experimental Activity 5 is used in this activity and is shown in Fig. 18.1. The nonlinearity (piecewise linear) property of this circuit is created by the two diodes connected in parallel with the capacitor C_2. A plot of the voltage across both capacitors as a function of charge would appear as shown in Fig. 18.2 and it should be noted that the reciprocal of the slope in a given region is the capacitance for that region. The diodes act as

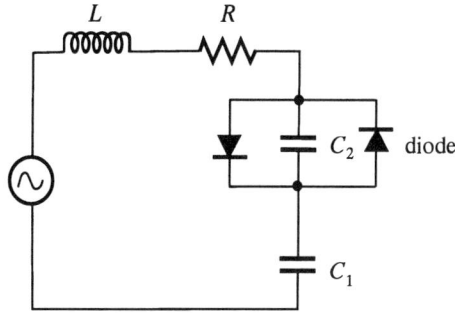

Figure 18.1: Circuit for producing a piecewise linear response.

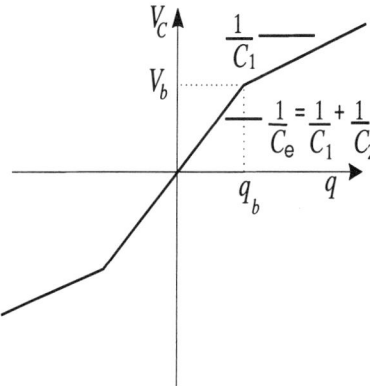

Figure 18.2: Voltage across both capacitors vs. charge.

open switches when the voltage is below some critical level ($V_b \approx 0.7$ volts) and act as electrical shorts to eliminate the capacitor C_2 when the voltage exceeds this critical value. When the diodes are acting as open switches both capacitors are in operation so the equivalent capacitance of the two capacitors in series is

$$C_e = \frac{C_1 C_2}{C_1 + C_2} \tag{18.1}$$

and the corresponding high resonance frequency is given by

$$\nu_{\text{high}} = \frac{1}{2\pi\sqrt{LC_e}}. \tag{18.2}$$

When the diodes are acting as shorts, C_2 is effectively removed from the circuit. Then $C_e = C_1$ and the low resonance frequency is given by

$$\nu_{\text{low}} = \frac{1}{2\pi\sqrt{LC_1}}. \tag{18.3}$$

For a "soft spring" oscillator the magnitude of the amplitude $|A|$ versus the frequency ω is as shown in Fig. 18.3. Notice that as the forcing amplitude F

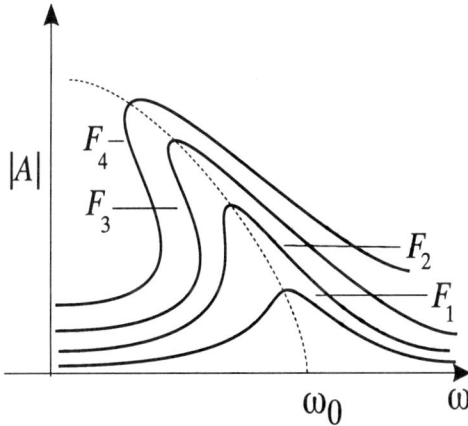

Figure 18.3: Amplitude $|A|$ vs. frequency (ω) for a "soft spring" oscillator.

increases $(F_4 > F_3 > F_2 > F_1)$ all the resonance frequency curves tilt to the left. This causes the respective resonance frequency (ω_0) curve (the dashed line) to tilt to the left also.

The piecewise linear function produces a similar tilting. The resonance curve for a small forcing amplitude (voltage) starts at ν_{high} and cannot go any lower than ν_{low}. A plot of the forcing amplitude as a function of resonance frequency is shown in Fig. 18.4 For comparison, a plot of the forcing amplitude–resonant

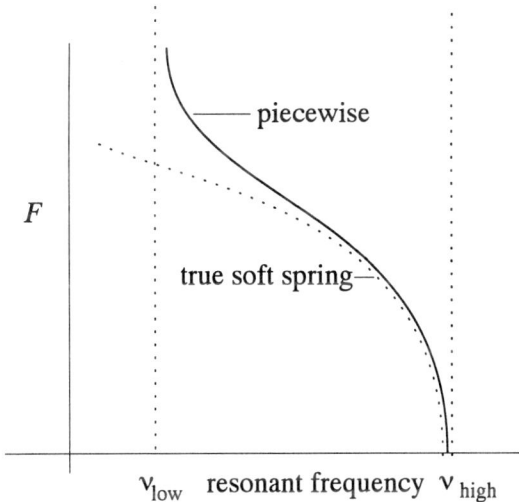

Figure 18.4: Forcing amplitude vs. resonance frequency.

frequency for a true "soft spring" oscillator is also shown on the graph. In this activity a plot similar to the solid curve in Fig. 18.4 is produced.

Procedure:

1. Construct the circuit as shown Figure 18.5.

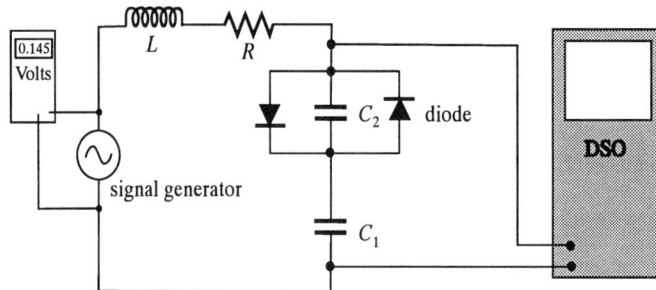

Figure 18.5: Piecewise linear circuit.

2. Use a signal generator with a low output impedance (50 Ω). The value of the inductance is not critical, but generally the larger the values the better. The 0.80 H, 67 Ω, Berkeley lab solenoid works well. Use capacitors that range from 1 μF to 0.047 μF. Keep C_1 larger than C_2 so that the piecewise bend is pronounced. A reasonable value for R is 50 Ω or lower. You want the circuit to have a large quality factor (Q) because a large Q means a sharp resonant spike. Remember that for small damping, $Q = \frac{\omega}{2\gamma}$. Here $2\gamma = \frac{R}{L}$ and $\omega = \frac{1}{\sqrt{LC_e}}$ and therefore $Q = \frac{\sqrt{L}}{R\sqrt{C_e}}$. To get this large Q you might wish to leave the external resistance R out of the circuit and rely only on the small intrinsic resistance of the wiring. However, if you do remove the external resistance, it makes it more difficult to keep the forcing amplitude constant as you near resonance. The reason for this is that the impedance of the external circuit gets smaller and smaller as you approach resonance, so if there was no external resistance and if the signal generator had no output impedance, the current in the circuit would be infinite. Signal generators are given an output impedance to prevent this catastrophe.

3. Using Eq. (18.2) and Eq. (18.3), calculate the two limiting resonant frequencies.

4. If the high resonance frequency is below 1000 Hz, you should be able to use a digital multimeter to check the signal generator's output voltage (RMS). If not, use an oscilloscope.

5. With the signal generator producing a low amplitude signal of around 0.10 volts (RMS), locate the circuit's resonant frequency. Do this by slowly increasing or decreasing the signal generator's frequency and watching the CRT trace (or meter) until a maximum amplitude is observed. After the resonant frequency has been located, note the forcing amplitude as it probably has changed a little.

6. Increase the forcing amplitude to 0.20 volts and repeat the above step.

7. Continue to increase the forcing amplitude in steps of 0.10 volts and repeat the above measurements.

8. When the forcing amplitude nears 1.0 volts you will be approaching the low frequency limit of this circuit. Decide for yourself when and where the last measurement should be made.

9. Construct a graph of the forcing amplitude as a function of the resonant frequency. Confirm the similarity and the difference between your curve and the theoretical curve for a "soft spring" oscillator?

Things to Investigate:

- Repeat the experiment with capacitors that have different values.

- Did you see any evidence of hysteresis as the frequency was varied? If you wish to modify the circuit to investigate hysteresis, see [FJB85].

- Simulate this circuit's performance with a SPICE program or with a similar program such as MICRO-CAP IV.

Experimental Activity 19

Nonlinear Resonance Curve: Magnetic.

Comment: This experimental activity should take less than 2 hours to complete.

Reference:

1. *Nonlinear Physics with Maple for Scientists and Engineers, 7.2.*

Object: To investigate and construct a nonlinear resonance response curve for the forced hard spring Duffing equation.

Theory: Nonlinear and linear resonance response curves are constructed by plotting the modulus of the steady state amplitude $|A|$ versus the driving frequency (ω). Figure 19.1 shows typical linear and nonlinear resonance response

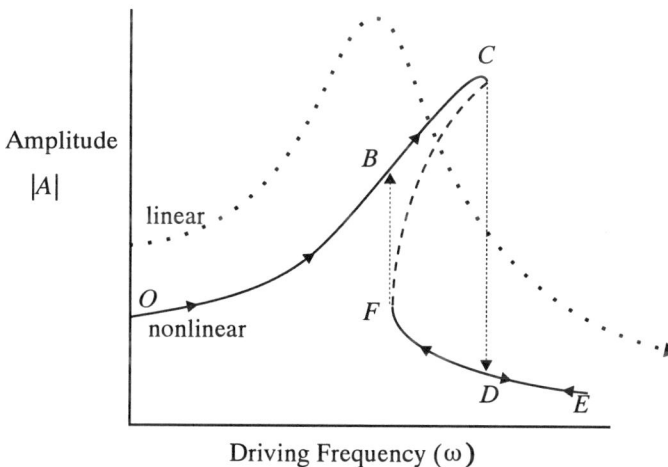

Figure 19.1: Linear and nonlinear resonance response curves.

curves. In the nonlinear case as the driving frequency is slowly increased from O, a point C is reached where the value for the amplitude undergoes a sudden drop from C to D and then continues to point E and beyond. However, if the frequency is slowly decreased from point E, the amplitude will undergo a quick jump from F to B and then will continue along the curve to O. The frequency at which the jump from F to B occurs is different than that of the drop frequency. The phenomenon of following different system paths when moving in different directions is known as hysteresis.

In this experiment a small cylindrical bar magnet will be suspended in an oscillating magnetic field. The torque (τ) between the magnetic moment (μ)

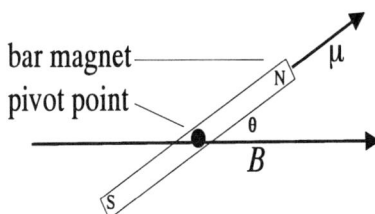

Figure 19.2: Magnet suspended in an external magnetic field.

and the vibrating magnetic field (B) provides the forcing function

$$\tau_f = \mu B \sin \theta. \tag{19.1}$$

The magnet is suspended by using VHS video recording tape, held under tension. The tape acts as a hard spring. It is assumed the tape provides a restoring torque similar in form to the equation

$$\tau_r = -\alpha \theta - \beta \theta^n \tag{19.2}$$

where n is approximately 3 and α and β are positive constants. The net torque is $\tau = I\ddot{\theta}$ where I is the moment of inertia of the cylindrical magnet. The vibrating magnetic field has the form $B = B_0 \cos(\omega t)$ where B_0 is the maximum value of the magnetic field. Combining the above equations, and taking $n = 3$, produces the nonlinear differential equation

$$I\ddot{\theta} + \alpha \theta + \beta \theta^3 = \mu B_0 \cos(\omega t) \sin \theta. \tag{19.3}$$

As the driving frequency ω is varied, hysteresis may occur. This is the subject of the present activity.

Procedure:

1. Construct the apparatus shown in Fig. 19.3. Place the magnet as close as possible to the solenoid or better yet at the center of the solenoid or Helmholtz coils. The magnet's supporting apparatus was built with a wooden Tinker Toy construction set. This supporting apparatus allows the tape's tension to be easily adjusted, and also allows the magnet to be

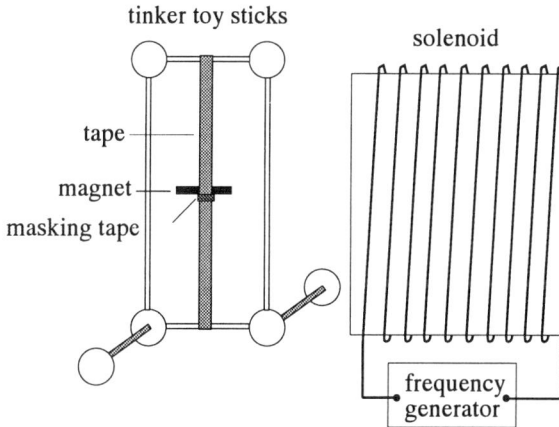

Figure 19.3: Solenoid, suspended magnet, and power supply.

placed correctly in the vibrating magnetic field. The magnet is held between a piece of doubled over VHS video tape. Use masking tape to pinch
the VHS tape just below where the magnet is to be supported. Adjust
the VHS tape's tension if the resulting oscillations are too small or large.
The solenoid that comes with the Berkeley Physics Experiments works
very well in this activity, but any large solenoid or Helmholtz coils could
be substituted. A cylindrical magnet about 5 cm long with a diameter of
0.50 cm works well, but any small strong bar magnet should work.

2. With the frequency generator set to produce a large (maximum) amplitude signal, start from a low frequency and slowly increase the driving
   frequency. Record the frequency where the amplitude undergoes a sudden
   drop.

3. Starting at a high frequency slowly reduce the driving frequency and record
   the frequency where the sudden jump occurs. Was hysteresis present?

4. Repeat the steps, but now record the values for the amplitude as the
   frequency is increased and decreased. Take two or three measurements
   past the drop and jump frequency. The amplitude can be measured by
   casting a shadow of the vibrating magnet on a wall. For a more accurate
   measurement of the amplitude, attach a small mirror to the magnet and
   reflect a laser beam off the mirror and onto a wall.

5. Construct an amplitude (magnitude) vs. frequency graph.

Things to Investigate:

- Repeat the activity using a different magnet (size or strength).

- Repeat the activity using a different tape tension.

- Design an experiment to find the relationship between the tape's restoring torque and the angular displacement.

- Calculate the magnetic moment (μ) by carrying out the following steps:

    1. Calculate the value of the external magnetic field (B_0);

    2. With the magnetic field held at this value, displace the magnet a small angle, say $10°$, from its equilibrium position;

    3. Release the magnet and measure the period (T) of its small vibrations;

    4. Develop and use the equation

    $$\mu = \frac{4\pi^2 I}{T^2 B_0}$$

    to calculate the magnetic moment (μ).

Experimental Activity 20

Subharmonic Response: Period Doubling

Comment: This experimental activity should take less than 1 hour to complete.

Reference:

1. *Nonlinear Physics with Maple for Scientists and Engineers, 7.3.*

Object: To investigate the phenomena of period doubling and chaotic behavior in a nonlinear mechanical oscillator.

Theory: One of the main indicators that a oscillating system might become chaotic is that it exhibits period doubling before exhibiting chaotic motion as a control parameter is changed. Figure 20.1 is a plot of a signal indicating

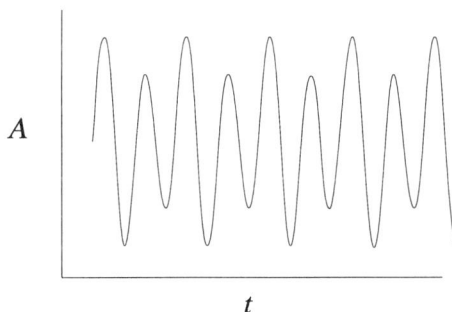

Figure 20.1: A signal showing a period 2 characteristic.

period doubling. The pattern repeats itself after two spikes (oscillations) and accordingly is known as a period 2 solution.

In this experiment, a small cylindrical bar magnet is suspended in a sinusoidal time varying magnetic field. The tape that suspends the magnet is assumed to provide a restoring force similar to that of a hard spring. As the angular displacement doubles, the restoring force increases by a factor larger than two.

The time-varying magnetic field provides a sinusoidal forcing function. (See Experimental Activity 19 dealing with Nonlinear Resonance Curves for a mathematical explanation and derivation of the nonlinear differential equation.)

Procedure:

1. Construct the following apparatus. Make sure the magnet is as close as possible to the solenoid or, better yet, at the center of the solenoid or Helmholtz coils.

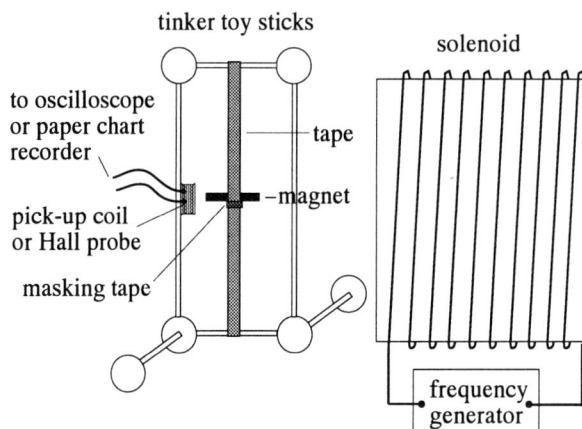

Figure 20.2: Solenoid, magnet, power supply, and pick-up coil.

2. The magnet's supporting apparatus can be built with a Tinker Toy construction set. This Tinker Toy apparatus allows for easy adjustment of the tape's tension by pushing the sticks in or out of their supporting blocks, and it also allows for building different sizes of supports so that the magnet can be correctly placed in the vibrating magnetic field. The magnet is held between a piece of doubled over VHS video tape. Use masking tape to pinch the VHS tape just below where the magnet is to be supported. If the magnet's amplitude is too small or too large, adjust the VHS tape's tension. For the driving solenoid, the solenoid that comes with the Berkeley physics experiments package works very well, but any large solenoid or Helmholtz coil could be used. A cylindrical magnet about 5 cm long with a diameter of 0.50 cm was used, but any small strong bar magnet should work. A small pick-up coil (CENCO's #79735-01T) or a Hall probe can be used to monitor the signal.

3. Connect the Hall probe or pick-up coil to a digital storage oscilloscope (DSO).

4. With the frequency generator set at suitable (usually maximum) power and at a low frequency, increase the driving frequency until the magnet's amplitude is just at its maximum.

5. Slowly reduce the frequency while watching the CRT trace for period doubling, quadrupling, etc.

6. If you are using an oscilloscope to examine the oscillations, draw sketches of the waveforms.

7. For a period 2 signal, measure the time between two large spikes, and the time between a large spike and a smaller spike. Using the driving frequency, calculate the driving period. Which of the measured times is equal to the driving period?

8. Try to adjust the driving frequency to produce a period 4 solution. Repeat the previous step.

Things to Investigate:

- Repeat the activity using a different magnet (size or strength).

- Repeat the activity using a different tape tension.

- Design an experiment to find the relationship between the tape's restoring torque and the angular displacement.

Experimental Activity 21

Diode: Period Doubling

Comment: This investigation should not take more than 1 hour to complete.

References:

1. *Nonlinear Physics with Maple for Scientists and Engineers, 7.3.*

2. [HJ93] Source of the circuit used in this activity.

3. [Bri87] Discussion of how diodes produce period doubling.

Object: To investigate the period doubling route to chaos.

Theory: Although one of the most common uses for a diode is to rectify AC current, in this activity the diode is used to produce a signal that exhibits period doubling and period quadrupling. Figure 21.1 shows the simple circuit used in this experiment. Initially there was some dispute of how a diode produces a

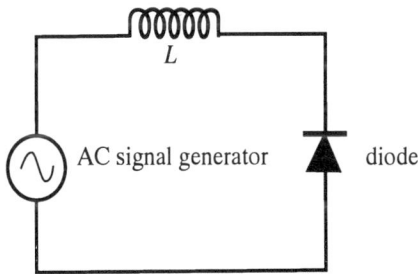

Figure 21.1: Diode circuit.

period doubling signal, but more recently a consensus has been reached. An expansion of the explanation given in [Bri87] and [HJ93] is now presented.

For the circuit shown in Fig 21.1, period doubling takes place if the frequency of the circuit is near the diode's resonance frequency. This frequency is determined by the diode's internal capacitance and the external inductance.

The resonance frequency of a diode does not remain constant because the capacitance of a diode changes as the strength of the applied signal across the diode changes, as shown in Figure 21.2. When the correct frequency set by the

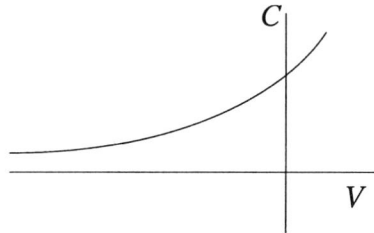

Figure 21.2: Capacitance vs. voltage for a diode.

AC signal generator is present, the amount of current flowing through the diode can affect the time it takes the diode to recover to its reverse bias equilibrium. The nonlinearity arises from the unrecombined charges that have crossed the diode's p–n junction when the diode is in the forward bias mode. These charges do not instantly recombine when the diode switches to the reverse bias mode, but diffuse back to their "home" region. Initially, these unrecombined charges act like a battery that produces a transient reverse bias current. This transient reverse bias current can be much larger than the reverse bias saturation current and even stranger, it may be larger that the forward bias current. To see how this large reverse bias current occurs, consider a signal generator wired in series with a resistance (R) and a diode. Assume the signal generator is producing a square wave of amplitude V_0. See Figure 21.3. When in the forward bias the

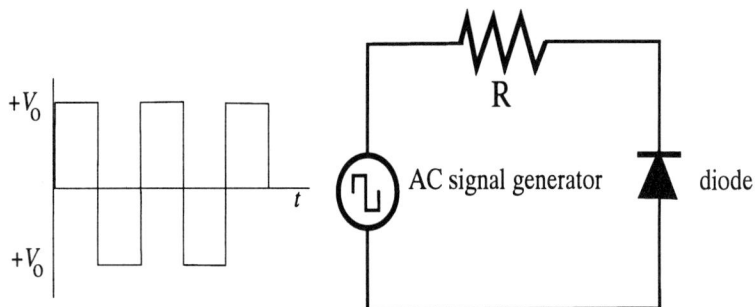

Figure 21.3: Square wave input for diode circuit

value for the current is given by $i = \frac{V_0 - V_d}{R}$. When the square wave instantly switches polarity, the diode is placed in the reverse bias mode. The trapped unrecombined charges maintain the forward diode potential of V_d, so the reverse current is given by $i = \frac{-V_0 - V_d}{R}$ which (momentarily) has a larger magnitude than the forward bias current. In fact, if V_0 is just a little larger than V_d, then this transient reverse bias current can be many times larger than the forward bias current. The trapped unrecombined charges act to delay the recovery of the diode. As the charges flow back across the junction the depletion region widens and the transient reverse bias current decreases until its value reaches the value of the normal reverse bias current. The larger the forward current, the greater the number of unrecombined charges and the longer it takes for the diode to reach its reverse bias equilibrium. Figure 21.4 shows a set of plots that indicate the behavior of the diode as it switches from forward to reverse bias. The first plot in Figure 21.4 shows the relationship between current and time.

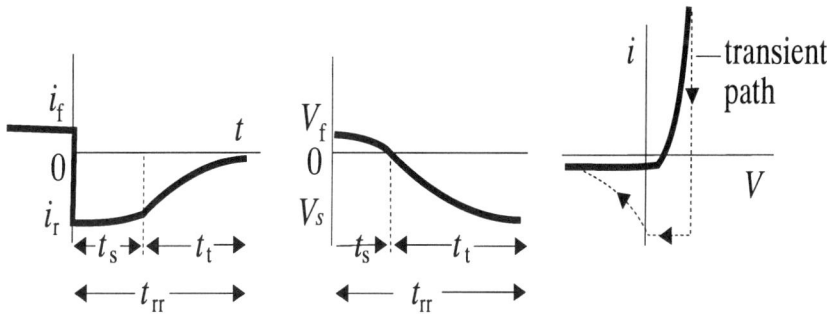

Figure 21.4: Curves for a diode switching from forward bias to reverse bias.

The symbol t_s is the storage time, the time that unrecombined charges are still providing a potential. The symbol t_t is the transition time, the time required for the large reverse bias current to return to the normal reverse bias current. The symbol t_{rr} is the total reverse recovery time. The second plot shows the relationship between the potential difference across the diode as a function of time. The third plot shows the plot of current vs. voltage for a normal diode with the transient current path overlaid upon it.

For certain frequencies, the circuit can switch back to its forward bias mode before the reverse bias equilibrium has been fully established. When this occurs, the behavior of the diode's next cycle depends on the parameters of the previous cycle. Each subsequent cycle could have a different initial condition. If this were to occur, chaos would result. To produce a period 2, period 4, etc., solution, it is necessary that the initial conditions are appropriately repeated.

Procedure:

1. Set up the apparatus as shown in Fig. 21.5. The diodes 1N914 or 1N4001 are inexpensive and work well.

2. The solenoid recommended for the PSSC apparatus works well as the inductor. It has very low resistance with an inductance of about 4 mH.

3. The signal strength and frequency of the generator is read from Channel 2. This signal acts as a reference for the signal displayed on Channel 1. The trace displayed on Channel 1 is the voltage-time signal for the diode. It is this signal which shows the period doubling.

4. For the diode shown, explore signal generator frequencies between 10 kHz and 90 kHz. For the first measurement, set the signal generator to a low

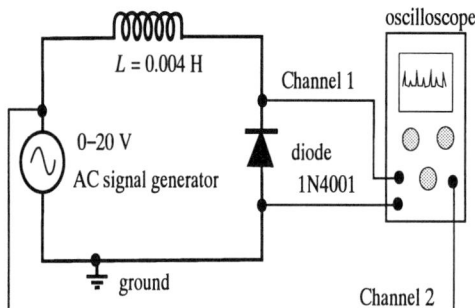

Figure 21.5: Period doubling circuit.

frequency and a low amplitude signal. With the generator producing a constant frequency slowly increase the strength (amplitude) of the signal. Watch Channel 1 on the oscilloscope for signs of period doubling. Figure 21.6 is an example of a period 2 signal. If no period doubling oc-

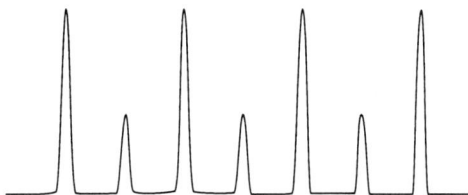

Figure 21.6: Period 2 signal.

curs reduce the signal generator's amplitude to zero, increase the signal generator's frequency by 5 kHz and slowly increase the signal's amplitude.

5. Repeat the above procedure until a period 2 signal occurs. Record the frequency and amplitude of the signal.

6. When period doubling is encountered, record the signal generator's frequency and amplitude (signal strength) from the trace shown on Channel 2.

7. Continue increasing the signal generator's amplitude to see if period 4 or even period 8 can be detected. When first encountered, record the signal generator's amplitude.
   ** When the authors performed this activity a period 4 signal was located at 50 kHz with an AC signal voltage of about 4 volts.**

8. For what range of frequencies on the signal generator does the period doubling sequence occur?

9. To produce phase plane portraits, switch the oscilloscope to its x–y setting. Sketch the CRT traces for period 2, 4, etc.

Things to Investigate:

- Use a larger inductance and repeat the above steps. Does this lower or raise the frequency at which period 2 starts?

- At a point where period 2 is just occurring, slowly insert a ferromagnetic core into the solenoid. Does the period 2 solution remain?

- When the oscilloscope is showing a period 2 or period 4 signal, set the oscilloscope to trigger on the largest pulse. Watch to see if the smaller pulse always occur at exactly the same place.

- If the strength of the signal is really the parameter that causes the non-linearity, speculate on why the period doubling sequence disappears when the signal strength is increased past a certain value.

Experimental Activity 22

Power Spectrum

Comment: This experimental activity should take less than 2 hours to complete.

Reference:

1. *Nonlinear Physics with Maple for Scientists and Engineers, 7.4.*

Object: To use a fast Fourier transform (FFT) to construct a power spectrum from experimental data.

Theory: A power spectrum is an important and useful diagnostic tool in trying to ascertain the frequency content of an oscillating time series. Section 7.4 of Reference 1, with its included Maple file 32 should be reviewed before attempting this activity.

One of the main precursors of the onset of chaotic motion is the appearance of an oscillation that exhibits period doubling, a typical plot being shown in Figure 22.1. Data can be lifted from time series plots so that a corresponding

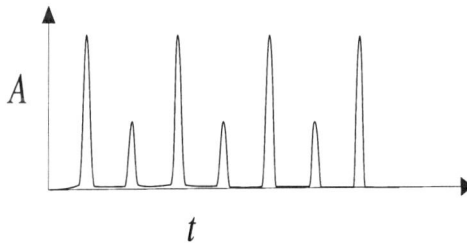

Figure 22.1: A signal showing a period 2 characteristic.

power spectrum can be constructed. For example, if the above graph contained a large number of peaks, and if the sampling data was collected correctly, a plot similar to the one shown in Fig. 22.2 would result. The tall peak is at the driving frequency and the smaller peak to the left of the tall peak is a subharmonic at half the driving frequency and thus indicates period doubling. The spike to

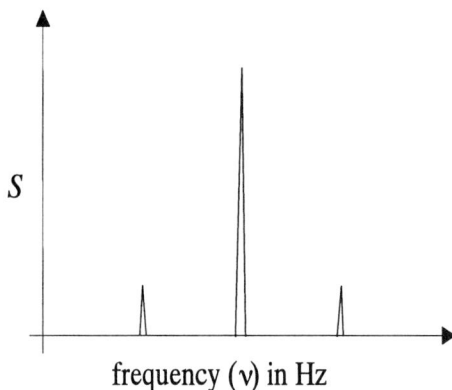

Figure 22.2: Power spectrum S for a period 2 solution.

the right of the center peak is called an ultrasubharmonic and is at three times
the frequency of the subharmonic peak. In this experiment a complicated time
series graph will be analyzed to produce a power spectrum.

Procedure:

1. If you have a hard copy of data from any previous activity, analyze that
   data and omit the following data collection steps.

2. If you do not have hard copy, construct the apparatus shown in Fig. 22.3.
   The details of construction and use may be found in Experimental Activity
   20.

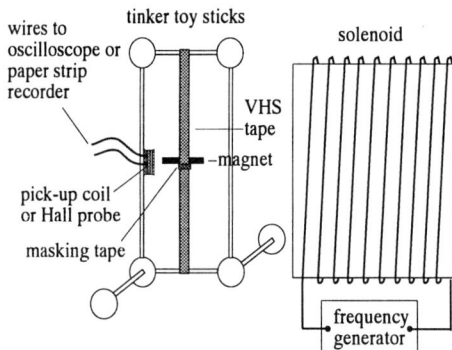

Figure 22.3: Solenoid, suspended magnet, power supply, and pick-up coil.

3. Connect the pick-up coil or a Hall probe to a paper strip recorder.

4. With the frequency generator set at a suitable (usually maximum) power
   level and at a low frequency, increase the driving frequency until the magnet's amplitude is just at its maximum. The magnet should vibrate not
   rotate.

5. Slowly reduce the frequency while watching the strip recorder's plot for examples of period doubling, tripling, etc. When a reproducible pattern is noticed, leave the signal generator at this frequency.

6. After some time and when the trace shows a large number (> 20) of similar oscillations, record the signal generator's driving frequency and the speed (cm/s) of the paper moving through the strip recorder.

7. Use the trace on the strip of paper to calculate the fundamental frequency and compare it with the frequency of the signal generator.

8. Select a suitable portion of the trace and subdivide it into equal time intervals. Each interval is equal to the sampling period (T_s). The total number of intervals must be equal to an integer multiple of a power of 2, i.e., 4, 8, 16, 32, 64, ... etc. For each marked point T_s, $2T_s$, $3T_s$, ..., record the amplitude of the wave at that specific spot. Experience has shown that 64 points are a minimum number to sample, but the more the merrier.

9. Maple file X22PWR.MWS calculates the power spectrum. Place the measured amplitudes into this file and run it. The file provides additional instructions to help you do this.

10. Does the power spectrum confirm the periodicity that you originally observed on the trace?

11. Locate the peak that corresponds to the fundamental frequency. Locate peaks to the left of the fundamental which indicate the occurrence of subharmonics. Identify these peaks by their frequencies.

12. Locate the peaks to the right of the fundamental frequency. Identify them in terms of multiples of the fundamental and subharmonic frequencies.

13. Repeat the procedure for twice as many data points.

Things to Investigate:

- Do a power spectrum of a stock market graph showing a stock's fluctuating value as a function of time.

- Do a power spectrum of some other type of data, such as weather or climatic effects, biological data, etc.

Experimental Activity 23

Entrainment and Quasiperiodicity

Comment: This investigation should not take more than 1 hour to complete.

References:

1. *Nonlinear Physics with Maple for Scientists and Engineers, 7.6.*

2. [HJ93] This article provided the idea for the following activity.

Object: To investigate entrainment and the quasiperiodic route to chaos.

Theory: As reported some three centuries ago by the Dutch physicist Huygens, two clocks on the same wall can become synchronized, the coupling being through the wall. This phenomena is called entrainment. A more modern and practical example of entrainment is in the use of an electronic periodic pacemaker to control the rhythm of the heart.

Quasiperiodicity occurs when neither frequency of the coupled oscillators wins out and the coupled system jumps from one frequency to the other.

This activity is a modification of Experimental Activity 21, which produced period doubling by using a diode wired in series with an inductor. In this experiment a diode is wired in series with an inductor and is then placed in parallel with another diode and inductor, as shown in Figure 23.1. Experimental Activity 21 gave a plausible explanation of how a diode in this configuration produces a nonlinearity that is required for period doubling behavior. Reviewing the theory from this activity might be useful.

Experimental Activity 21 showed that a single branch of this circuit can produce period doubling when operating at the correct frequency and signal strength. When in tandem, the parallel branches modulate the behavior of each other. As the signal strength increases and as each branch begins to exhibit period doubling, the frequencies remain close but incommensurate (the ratio of the two frequencies is irrational), but as the signal strength is increased even further the frequencies can phase lock or become entrained. At even higher

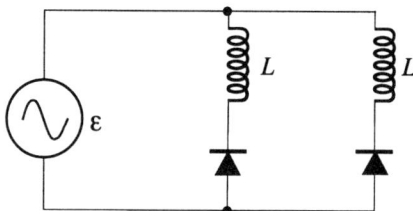

Figure 23.1: Circuit to exhibit entrainment and quasi-periodicity.

signal strengths a quasiperiodicity arises as the signal moves back and forth from one inductor's preferred frequency to the other inductor's preferred frequency.

Procedure:

1. Set up the apparatus as shown in Fig. 23.2. For the diodes, use any general purpose diode, e.g., 1N914 or 1N4001.

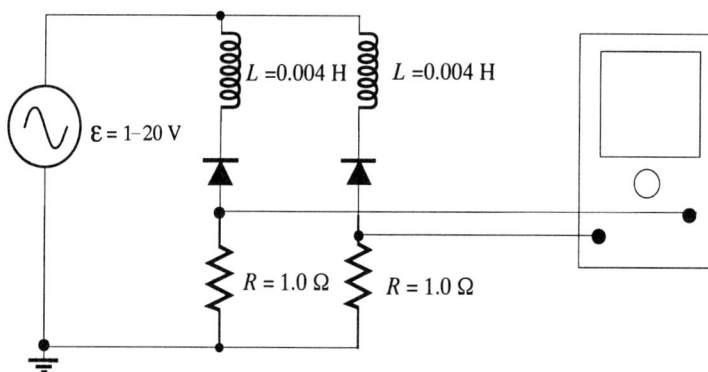

Figure 23.2: Experimental set-up.

2. The solenoid recommended for the PSSC apparatus works well as the inductor. It has very low resistance with an inductance of about 4 mH.

3. The two signal strengths are monitored across each of the resistors and fed into channels 1 and 2 on the oscilloscope.

4. If Experimental Activity 21 has been completed, the signal strengths and frequencies needed to produce period 2 are known. If not, frequencies between 10 kHz and 90 kHz should produce the desired effects.

5. Set the signal generator so that it produces a low amplitude signal. With the generator producing a constant frequency, slowly increase the strength (amplitude) of the signal. Watch the oscilloscope in its dual channel mode for signs of period doubling in each branch of the circuit. Switch to the

x–y setting to observe the coupling of the two signals. Figure 23.3 is a schematic representation of one of the shapes that can be observed in the x-y mode. The figure, referred in the cited literature as the Klein bottle

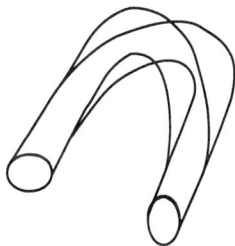

Figure 23.3: The folded torus known as the Klein bottle condom.

condom, is an example of quasiperiodicity. Switching the oscilloscope back and forth from its x–y setting to its dual–channel mode may help you decipher what the individual and coupled signals are doing.

6. Alter the amplitude or frequency and search for evidence of entrainment.

7. Continue increasing the amplitude of the signal to see if you can detect when chaos begins.

Things to Investigate:

- Use one solenoid with a little smaller inductance than other and see if you can still make the frequencies entrain.

- Try using the other diode type mentioned in the first step of the procedure. How does this change affect the data?

Experimental Activity 24

Quasiperiodicity: Neon Bulb

Comment: This investigation should not take more than 1 hour to complete.

References:

1. *Nonlinear Physics with Maple for Scientists and Engineers*, 7.6.

2. [Moo92] The idea for the circuit in this activity was found in this book.

3. Experimental Activity 13 dealing with a neon bulb relaxation oscillation.

Object: To investigate the various ways that two neon bulb relaxation oscillators can interact to produce quasiperiodic and chaotic flashes of light.

Theory: As reported some three centuries ago by the Dutch physicist Huygens, two clocks on the same wall can become synchronized, the coupling being through the wall. This phenomena is called entrainment. A more modern and practical example of entrainment is in the use of an electronic periodic pacemaker to control the rhythm of the heart.

Quasiperiodicity occurs when neither frequency of the coupled oscillators wins out and the coupled system jumps from one frequency to the other. This activity investigates the above phenomena using coupled oscillators produced by two neon glow lamps.

Figure 24.1 is a diagram of a neon glow lamp. The small bulb is about

Figure 24.1: A Ne-2 neon glow lamp.

1.0 cm long and contains two vertical metal electrodes which are separated by about 2 mm. The electrodes are surrounded by a low pressure neon gas. When a potential difference is applied across the electrodes, it creates an electric field that accelerates charged ions and electrons that exist inside the bulb. These stray ions and electrons are always present due to cosmic radiation and natural radioactivity. If the electric field is large enough, it can accelerate the electrons to a speed that is sufficient to create an avalanche of electrons. For example, when an electron collides with a neon atom, a second electron is ejected and now there are two electrons available for the next collision, which results in four electrons available for the next collision and so on—the avalanche has begun. The negative electrode attracts the positive neon ions and the electrons leaving the electrode recombine with the positive neon ions to produce a soft glow.

Accordingly, a neon glow lamp is considered to have a very high resistance until the correct potential difference exists across the electrodes. At the critical potential difference—the firing voltage (V_f)—the presence of a seed electron creates an avalanche of electrons and the bulb begins to conduct. This conduction reveals itself by making the lamp glow. Once the neon bulb is glowing, its resistance decreases dramatically, so a current limiting resistance is usually placed in series with the bulb. The current limiting resistance is normally selected to have a value that permits the bulb to stay glowing without burning out or turning off. As the current through the bulb increases and the potential drop across the bulb decreases, the potential drop across the resistance increases. Accordingly, three operating conditions are possible:

1. If the current limiting resistance is very large, the bulb turns off;

2. If the current limiting resistance is too small, the current continues to increase and the bulb burns out;

3. If the current limiting resistance is correctly selected, the bulb remains lit.

Because neon bulbs consume very little power and have a very long life, they are used to indicate if a device such as a power supply is turned on. For a more complete description of how the neon glow lamp functions see the theory given in Experimental Activity 13.

Procedure:

The voltages used in this activity can produce severe or even fatal electrical shocks, so please be careful. Always turn off the circuit and let the capacitors discharge before making changes in the circuit.

1. Build the circuit shown in Fig. 24.2. The variable DC power source (ε) should be able to produce values from 0 to 400 volts. The resistors marked R_1 and R_2 should each be 1 MΩ. The capacitors C_1 and C_2 should have identical values around 0.47 μF. The capacitors should be rated at least half the maximum voltage used, so 250 volts is a safe value. The neon glow lamp or bulb is of the class Ne-2. These bulbs have a variety of extinction and firing voltages but the firing voltage is usually between 70 and 110

volts and the extinction voltage between 50 and 70 volts. Bulbs with lower firing voltages are preferable, because this permits power supplies of lower voltage to be used.

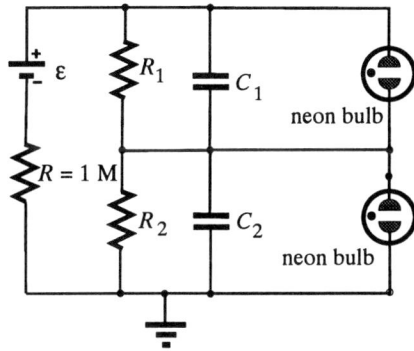

Figure 24.2: Coupled neon bulb circuit.

2. Slowly increase the value of the source voltage until the lights just begin to flash.

3. Slowly increase or decrease the source voltage to produce flashes that might indicate that chaotic oscillations could be imminent. Hint: Do the flashes indicate one or more of the following properties: quasiperiodicity, entrainment (mode-locking), period doubling, or intermittency?

4. Try to adjust the source voltage to produce chaos.

Things to Investigate:

- Replace R_1 with a variable resistor that differs from R_2 by 5% to 10%. Repeat the above steps.

- Add a third circuit to make three coupled oscillators.

Experimental Activity 25

Chua's Butterfly

Comment: This investigation should not take more than 3 hours to complete.

References:

1. *Nonlinear Physics with Maple for Scientists and Engineers, 7.7.*

2. [HJ93] This article is the source of the circuit used in this activity.

Object: To produce a chaos-exhibiting double-scroll strange attractor similar to the Lorenz attractor.

Theory: In this activity, an op-amp circuit is used to produce a negative resistance $(-r)$ similar to that encountered in previous activities, but with one main difference. The negative resistance is made piecewise linear so that the current–voltage line resembles that shown in Fig. 25.1. The negative resistance

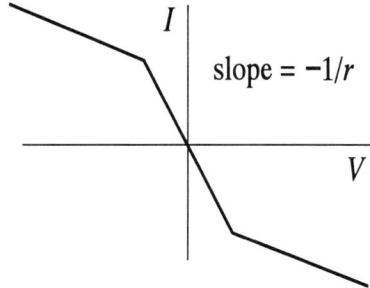

Figure 25.1: Piecewise linear negative resistance.

allows an oscillating signal to occur and its piecewise linear nature allows the system to become chaotic.

An equivalent circuit to the one used in this activity is shown in Fig. 25.2. Although the purpose here is mainly to observe the period doubling route to

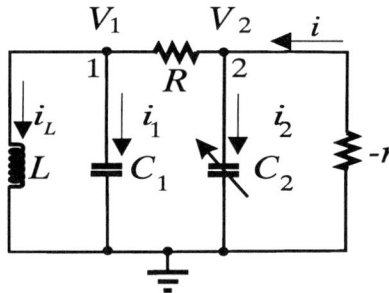

Figure 25.2: Equivalent circuit for state variable analysis.

chaos produced by this circuit, the equations that govern how this circuit functions are now developed using the state variables method for analyzing circuits. The state variables method, although useful in all circuits, is particularly useful when applied to nonlinear systems. This method permits the system to be studied by locating the stationary points and identifying the nature of the trajectories near these points. The process starts by identifying the number of state variables needed to describe the system. In electrical systems, the common state variables are the voltage drops across the capacitors and the currents flowing through the inductors. It should be noted that for any one system there is usually more than one set of state variables which will describe the system and that no one set of equations takes a mathematical precedence over the others.

When working with state variables, a few simple rules can be used to help set up the equations:

1. Identify the smallest number of variables that are needed to solve the system. The minimum number is equal to the number of degrees of freedom;

2. A rule of thumb is that the number of degrees of freedom is equal to the number of primary loops. This is the same idea as identifying the degrees of freedom in a mechanical system;

3. The number of equations that must be constructed is equal to the number of state variables;

4. The left side of the equations should only contain first-order time derivatives of one of the state variables;

5. No derivatives should appear on the right side of any of the equal signs;

6. The right side of the set of equations contains only the state variables, their coefficients, and the forcing functions if any are present,

7. Kirchhoff's voltage and current rules are used to set up the system of equations.

As an example of how the method of state variables is used, consider the circuit shown in Fig. 25.2. The circuit has three primary loops. Therefore it has three degrees of freedom, and a minimum of three state variables are required to solve the system. This requires three equations. In accordance with the standard convention, the variables for this circuit are chosen to be V_1 (the potential drop across C_1), V_2 (the drop across C_2), and i_L (the current through L). The sum of the currents into junction 2 must equal the sum of the currents out of junction 2, which produces

$$i = C_2 \frac{dV_2}{dt} + \frac{V_2 - V_1}{R}. \tag{25.1}$$

Assuming that the current through the negative resistor is some function of the voltage at nodal point 2, $i = f(V_2)$ and rearranging Eq. (25.1) gives

$$\frac{dV_2}{dt} = \frac{V_1 - V_2}{C_2 R} + \frac{f(V_2)}{C_2}. \tag{25.2}$$

Analysis at nodal point 1 gives

$$\frac{V_2 - V_1}{R} = i_L + i_1. \tag{25.3}$$

Since $q_1 = C_1 V_1$, then $i_1 = C_1 \frac{dV_1}{dt}$ and Eq. (25.3) becomes

$$\frac{dV_1}{dt} = \frac{V_2 - V_1}{C_1 R} - \frac{i_L}{C_1}. \tag{25.4}$$

Finally, by inspection of the circuit, the third equation is written as

$$\frac{di_L}{dt} = \frac{V_1}{L}. \tag{25.5}$$

Note that the system of equations is now complete because we have three equations in three unknowns. The student should note the similarity of this approach to phase space analysis. The student is encouraged to write a Maple file to solve this system of equations. The student will have to make use of Maple's piecewise capability. A piecewise example is provided in Experimental Activity 5.

As discussed in the accompanying text (e.g., Section 2.7), when one has a coupled system of three first-order nonlinear differential equations, chaos is possible. The investigation of chaotic behavior produced by the circuit which is governed by the above equations is the purpose of this activity.

Procedure:

1. Wire the circuit shown in Fig. 25.3. Notice the two diodes which create the piecewise function. These diodes act as switches which do not turn on until a potential drop of approximately 1.7 volts appears across the ground and the bottom wire on the diagram. When the diodes turn on, the 3.6 k resistors become part of the 2000 Ω negative resistance.

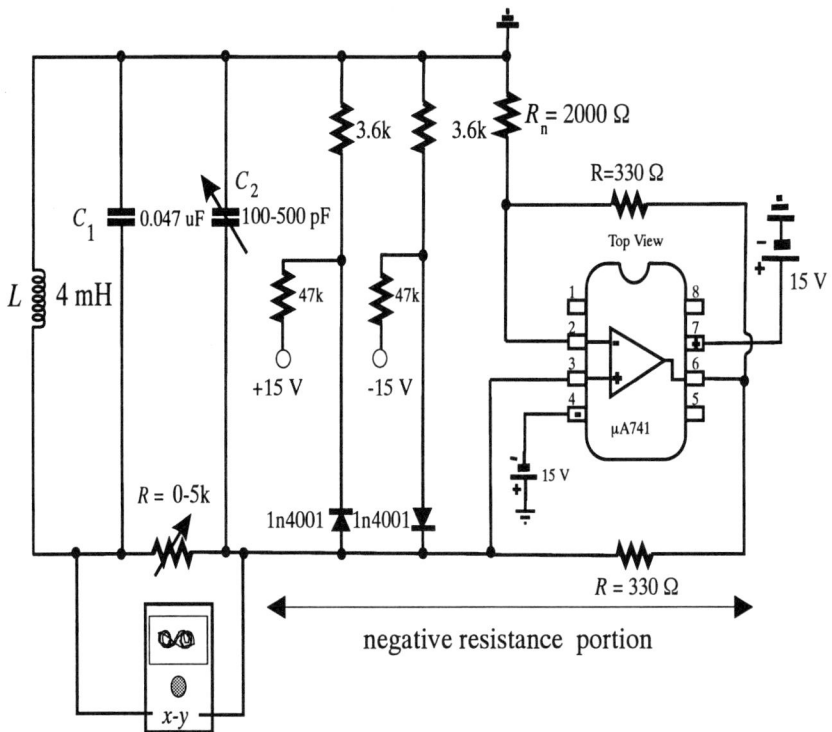

Figure 25.3: Chua's circuit to produce a double scroll strange attractor.

2. The PSSC solenoid used in so many of this manual's activities works well as the inductor. It has very low resistance with an inductance of about 4 mH.

3. A 0–5.0 kΩ analog variable resistor seemed to work better than a digital dial type.

4. The signal strength is read across the variable external resistor. This resistor is considered to be the positive resistor R in the equivalent circuit Fig. 25.2. To see the double-scroll strange attractor, the CRT must be set to its x–y setting.

5. Slowly reduce the resistance of the variable resistor from its maximum value to a point close to the initial negative resistance (2000 ohms in the circuit shown here). When this is done, a sharp jump occurs from the initial closed loop to a loop indicating period doubling (the trajectories cross), quadrupling, etc. This circuit is very sensitive to small changes in resistance, so carefully explore the critical region.

6. With finer tuning of the variable resistance, the jump to chaos is easily identified by the appearance of the double scroll. If you have difficulty producing a double scroll pattern, try changing the value of the variable capacitor.

7. When in the double scroll mode, switch the oscilloscope from its x–y setting to its dual channel setting. This should permit you to reproduce the time series plots shown in Fig. 7.29 and 8.25 of the accompanying text. These plots show quasiperiodic and chaotic behavior, respectively. Can you see the similarities?

Things to Investigate:

- Try to produce a chaotic signal using solenoids with larger and smaller inductances.

- Search the Internet for references to Chua's circuit. You will be surprised at how many references are cited and the number of ways the circuit is used.

- If you can produce a signal in the audio range (100–15,000 Hz), connect the output to an amplifier and then to a speaker. Listen to the chaos!

- Try to produce asymmetries in the piecewise function by changing one of the 3.6 kΩ resistors to a smaller or larger value. An alternate method of producing asymmetries is to reduce one of the diode's offset voltages from 15 V to some lower value.

Experimental Activity 26

Route to Chaos

The Route to Chaos

Comment: This investigation should not take more than 2 hours to complete.

Reference:

1. *Nonlinear Physics with Maple for Scientists and Engineers, 8.4.*

2. [Bri87] This article discusses a similar experiment.

3. [Chi79] Gives a parameter which can be used to decide if the motion in this activity is chaotic.

Object: To investigate forced nonlinear oscillatory motion.

Theory: A cylindrical bar magnet which is free to rotate horizontally about a vertical axis which is perpendicular to its horizontal longitudinal axis is placed in an external magnetic field (B) as shown in Fig. 26.1. In this configuration

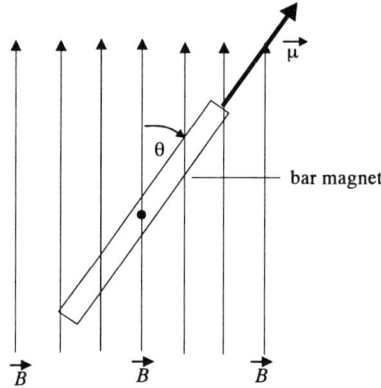

Figure 26.1: Bar magnet in time varying magnetic field.

the magnet will experience a restoring torque (τ) given by

$$\tau = -\mu B \sin \theta.$$ (26.1)

where μ represents the magnetic moment of the magnet and θ represents the angle between the magnetic field (B) and the magnetic moment. The negative sign in Eq. (26.1) indicates the restoring torque is in the opposite direction to the angular displacement (θ). If the magnitude of the magnetic field is a sinusoidal function of the time (t), then B varies according to $B = B_0 \cos(\omega t)$. Ignoring damping and the small restoring force exerted by the earth, Newton's second law $\tau = I\ddot{\theta}$ gives

$$\ddot{\theta} = -\frac{\mu B_0}{I} \cos(\omega t) \sin \theta,$$ (26.2)

where I is the moment of inertia of the magnet. If the length ℓ of the cylindrical bar magnet (mass m) is much larger than the end radius, then the magnet's moment of inertia is approximately equal to

$$I = \frac{1}{12} m\ell^2.$$ (26.3)

By setting $\tau = \omega t$, Eq. (26.2) can be put into the form

$$\ddot{\theta}(\tau) = -\frac{s^2}{2} \cos \tau \sin \theta$$ (26.4)

where $s = \sqrt{\frac{2\mu B_0}{I\omega^2}}$ plays the role of a control parameter.

This activity investigates motion of the bar magnet governed by Eq. (26.2) or Eq. (26.4). The motion exhibits a variety of nonlinear phenomena, including intermittency and chaos.

Procedure:

1. Set up the apparatus as shown in Fig. 26.2.

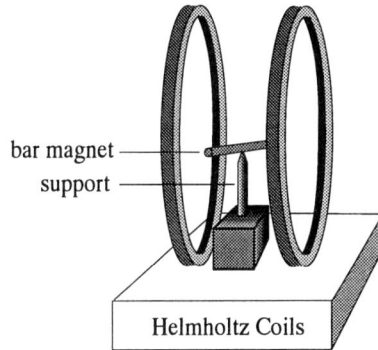

Figure 26.2: Experimental apparatus.

2. If the motion is to be observed without measurements, instead of a bar magnet use a standard classroom demonstration dipping needle or a demonstration compass. If measurements are to made, then a stronger magnet must be used and a frictionless method of suspending the magnet must be found. We rely on your ingenuity. One approach is to attach two neodymium magnets to a demonstration compass in a teeter-totter fashion to maintain the horizontal balance.

3. To monitor the signal use a Hall probe or a small pick-up coil connected to a chart recorder or to a digital storage oscilloscope (DSO).

4. With the signal generator strength set at maximum, alter the driving frequency until the magnet is set into a rotating motion.

5. Reduce the frequency until the motion appears chaotic.

6. What is the value for the critical driving frequency where the rotational motion becomes chaotic?

7. Change the strength of the driving signal and driving frequency. See if you can find or locate regions of intermittency. Intermittency is when the motion remains in one periodic state for some time before leaping to another periodic state. The decreasing length of time spent in any given periodic state is a measure of approaching chaos.

8. Keeping the external magnetic field constant, measure the period T of small oscillations around the equilibrium point.

9. Measure the mass m and length ℓ of the oscillating magnet. Calculate the magnet's moment of inertia I.

10. Calculate the value for the magnetic field B_0. Use the equation given for the Helmholtz coils, usually written on the apparatus. If a solenoid is used, calculate the strength of its magnetic field using standard textbook equations.

11. Calculate the magnetic moment (μ) of the magnet using

$$\mu = \frac{4\pi^2 I}{T^2 B_0}.$$

12. Write a Maple file that solves numerically Eq. (26.2). For a given frequency ω, compare the $\theta(t)$ predicted by the file with that observed experimentally.

13. Calculate the parameter s using the value found for the critical driving frequency. It has been predicted [Chi79] that chaotic motion occurs when $s > 1$. How does your value of s compare with the theoretical prediction?

14. Compare the frequency needed to efficiently pump the magnet from rest to a large amplitude with the frequency for small oscillations.

Things to Investigate:

- Modify Maple file 11 provided with the accompanying text, and investigate the Poincaré sections produced for Eq. (26.4) for different s values.

- Rewrite Eq. (26.2) to include the restoring force of the earth's magnetic field. Write a Maple program to investigate the difference this makes in the predicted motion. Is the ignoring of the earth's magnetic field justified?

- Repeat the steps in the procedure with the magnet suspended so it can rotate vertically.

- Suspend the magnet so it can rotate vertically, but has its center of mass initially below the fulcrum. This is now analogous to a compound pendulum. Try pumping this pendulum and study its motion. Write the differential equation that describes this situation and compare its predictions with the observations.

- Analyze the signal using a fast Fourier transform (FFT) to produce a power spectrum. Does the power spectrum indicate possible chaos?

- Add a damping term to Eq. (26.2) and then write a Maple file to explore and compare the experimental behavior with the behavior predicted by Maple.

Experimental Activity 27

Driven Spin Toy

Comment: This investigation should not take more than 1 hour to complete.

Reference:

1. *Nonlinear Physics with Maple for Scientists and Engineers, 8.4.*

Object: To investigate forced nonlinear oscillatory motion.

Theory: This activity uses an inexpensive toy, sold under the name Revolution: The World's Most Efficient Spinning Device, as a black box oscillator[1]. Figure 27.1 is a diagram of the device. The movable cylinder and base of this

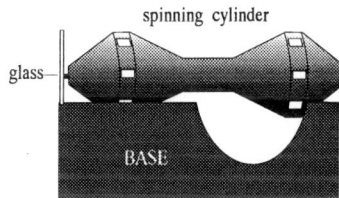

Figure 27.1: The spin toy oscillator.

toy contain magnets. These magnets levitate and hold the cylinder against the fixed glass end of the base. If the cylinder is placed in an oscillating magnetic field, the magnetic field acts as a forcing function. With the addition of an oscillating and pumping external magnetic field, the cylinder can be made to exhibit period doubling and chaotic motion. This investigation is mainly qualitative in nature. The quantitative portion of this activity explores the forcing frequencies and amplitudes required to produce chaotic motion.

Procedure:

1. Set up the following apparatus. The solenoid that comes with the Berkeley Lab course works well. The power supply (function generator) should be

---

[1] Arbor Scientific, P.O. Box 2750, Ann Arbor, MI 48106-2750, phone 1-800-367-6695.

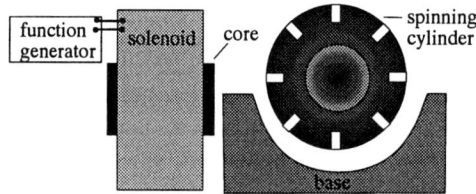

Figure 27.2: The forced oscillator.

able to be adjusted to produce sine waves of 0.1 to 1.0 Hz. The core is a laminated piece of iron that comes with a demonstration transformer set.

2. The spinning cylinder has white marks placed at 45° around its circumference. These marks can be used to measure the angular amplitude of the oscillations.

3. Measure the period T of the oscillating cylinder for a very small initial amplitude. Calculate the natural frequency ($\nu = \frac{1}{T}$) and the natural angular frequency (ω_0) for the spinning cylinder. This gives an approximate value for the pumping frequency.

4. With the cylinder initially at rest, and with the external magnetic field in resonance with the small oscillation frequency, observe the motion of the cylinder. At this pumping frequency, is it possible to make the cylinder complete a full revolution? Explain.

5. With the driving frequency still equal to the small angle frequency, give the cylinder a spin with your fingers and observe the resulting motion. Does the motion finally settle into the same pattern as that produced when the cylinder was initially at rest?

6. With the cylinder initially at rest, find a pumping frequency that makes the cylinder oscillate with a large amplitude. (Hint: A pendulum (a swing) is more efficiently pumped with a frequency nearly double that of its natural frequency.) Observe the motion to see if period doubling occurs. Be careful. Small frequency increases can make the cylinder rotate and chaotic motion might be the result.

7. Give the cylinder an initial spin with your fingers. With the same pumping frequency as that found in the previous step, observe the motion to see if it is qualitatively different from the previous motion.

8. Is the motion chaotic? What tests or measurements would have to be performed to see if the motion is chaotic?

9. When the cylinder is oscillating, change the driving amplitude or the driving frequency. Explore the effect this has on the cylinder.

10. Check the validity of your results by modifying the numerical solving section of Maple file X01SPI.MWS to construct a plot of your results.

Experimental Activity 28

Mapping

Comment: This investigation should not take more than 1 hour to complete.

References:

1. *Nonlinear Physics with Maple for Scientists and Engineers, 8.9.*

2. [DH91] This article contains a mapping procedure similar to the one in this activity.

Object: To construct a map for a forced nonlinear oscillator.

Theory: This activity uses commercially available toys, two of which are shown in Fig. 28.1. In the bottom supporting structure of each toy is a transistor circuit that is turned on every time a swinging magnet passes over it. The magnet moving over the top of the supporting base induces a current in a small coil hidden inside the base. This induced current when applied to the base terminal of the transistor momentarily turns the transistor on. At this

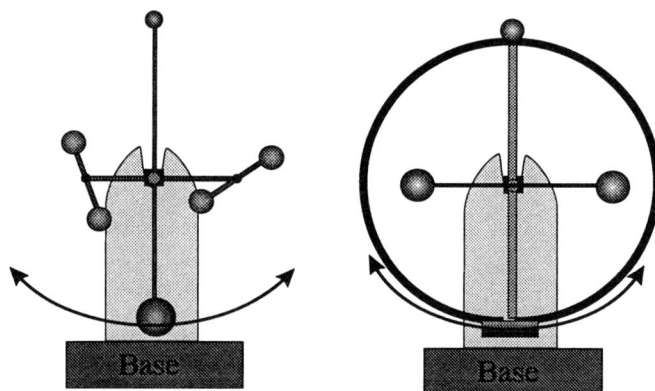

Figure 28.1: Original toys.

point a much larger current flows through a larger solenoid which is used to produce a repelling magnetic field. In this manner the swinging magnet is given a sharp push when it passes over the base of the toy. The time interval between successive pushes is measured. A map is created by plotting one time interval against the next time interval.

If the toys are left as pictured in Fig. 28.1 the pumping is very regular, but the rotation of the smaller pendulums is chaotic. To make the main rotation less periodic and to permit full rotations, the smaller pendulums are removed. (Sometimes a counter balancing object is needed to make the toy complete a full rotation.) Fig. 28.2. shows the modified toys. (If the toys you buy are

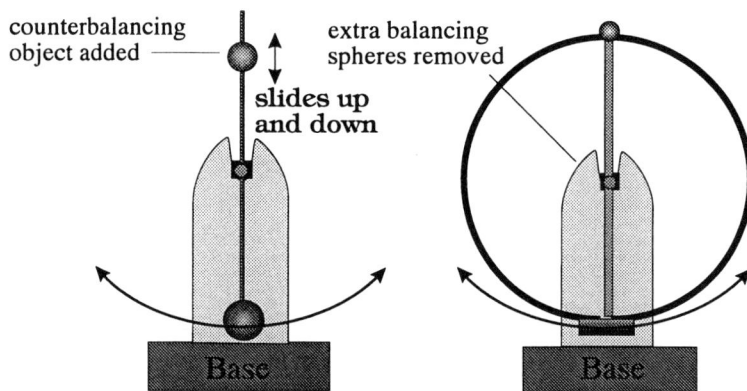

Figure 28.2: Modified toys.

different than those shown, then we rely on your ability to modify them so they will make full rotations.)

Procedure

   1. Set up the apparatus as shown in Fig. 28.3. This toy comes in a number of

versions and sizes. The smaller versions are easier to pump. (The authors used two toys, both of which had a stem height or circular diameter of 15 cm. If the smaller versions can not be found, then drive the toys with two 9 V batteries connected in series.) These toys can be purchased in many novelty shops. The counterbalancing object may be a rubber stopper with

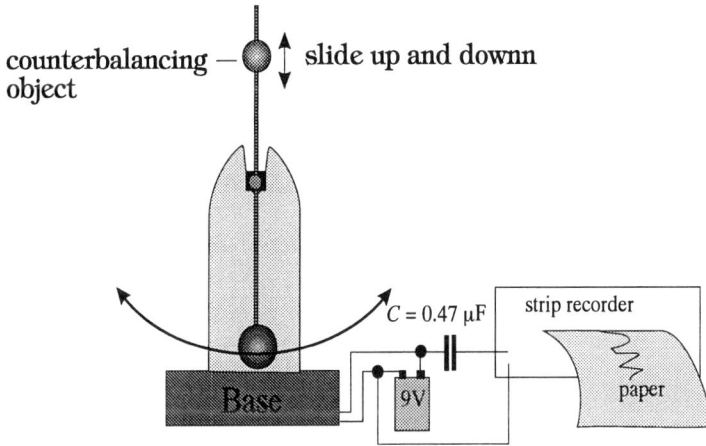

Figure 28.3: One of the toys used as a forced oscillator.

a hole drilled through its center or even the lead weights normally used on fish lines.

2. Remove, but do not disconnect, the 9 volt battery from the base of the toy. Attach two additional wires to the terminals of the battery and connect as shown in the diagram. The capacitor is essential and is used to block the 9 volt DC signal, but lets the pulsed signal pass.

3. With your hand give the toy a small push off its equilibrium point.

4. Adjust the movable counterbalancing object up or down until an interesting nonlinear rotational motion occurs.

5. Turn on the strip recorder and note the speed of the chart paper.

6. From the paper strip, measure the time intervals, Δt_1, Δt_2, Δt_3, ... between pushes. An alternate method for measuring the pumping times is to use a computer to monitor the times.

7. Create a map by placing on the ordinate Δt_{n+1} and on the abscissa Δt_n. Input the data into the provided Maple file X28MAP.MWS to see the map and explore various plotting configurations.

8. What does the plot tell you about the motion? Is there any evidence for a period doubling route to chaos? How can you tell if the motion is chaotic?

Things to Investigate:

- Interesting maps and very strange attractors were produced by collecting a very large number of points. This was done by using commercially available software to record the data. You might wish to do the same.

- Plot the amplitude of the strip recorder's signal against the time.

- Compare the strange attractors produced by this activity with those produced in the dripping faucet article [DH91].

Bibliography

[BN92]   R. Boylestad and L. Nashelsky. *Electronic Devices & Circuit Theory.* Prentice Hall, New Jersey, fifth edition, 1992. See pages 807–812.

[Bri87]   K. Briggs. Simple experiments in chaotic dynamics. *Am. J. Phys.,* **55**(12):1083–1089, December 1987. The idea for using a steel tape to model the Duffing equation was found in this article.

[Chi79]   B.V. Chirikov. *Phys. Rep.,* **52**:265, 1979.

[Cro95]   A. Cromer. Many oscillations of a rigid rod. *Am. J. Phys.,* **63**(2):112–121, February 1995. This article is highly recommended. It provides ideas for supporting an oscillating rod, diagrams illustrating different pendulum configurations and the equations for calculating their periods, and suggestions for additional and more complex investigations of rigid rod oscillations.

[DFGJ91] B. Duchesne, C. Fischer, C. Gray, and K. Jeffrey. Chaos in the motion of an inverted pendulum: An undergraduate laboratory experiment. *Am. J. Phys.,* **59**(11):987–992, November 1991. This article delves deeper into the theory of the inverted pendulum than does the Briggs' article.

[DH91]   K. Dreyer and F. Hickey. The route to chaos in a dripping water faucet. *Am. J. Phys.,* **59**(7):619–627, July 1991. This article contains maps produced by dripping faucets.

[DW83]   W.F. Drish and W.J. Wild. Numerical solutions of Van der Pol's equation. *Am. J. Phys.,* **51**(5):439–445, May 1983. This article contains values for the coefficients of the VdP equation that might be used by Maple to reproduce the article's plots. The article also contains a discussion of a nonlinear electric circuit that was proposed for use with elevators.

[FJB85]   E.L.M. Flerackers, H.J. Janssen, and L. Beerden. Piecewise linear anharmonic lrc circuit for demonstrating "soft" and "hard" spring nonlinear resonant behavior. *Am. J. Phys.,* **53**(5):574–577, June 1985.

[For87]   L.R. Fortney. *Principles of Electronics: Analog & Digital.* Harcourt Brace Jovanovich, Toronto, first edition, 1987. See page 308.

[HCL76]   P. Horn, T. Carruthers, and M. Long. Threshold instabilities in non-
          linear self-excited oscillators. *Phys. Rev. A.*, **14**(2):833–839, August
          1976. The original equation for the Wien bridge was found in this
          article.

[HJ93]   E. Hunt and G. Johnson. Keeping chaos at bay. *IEEE Spectrum*,
          pages 32–36, November 1993. This article contains a number of non-
          linear circuits.

[HN84]   Y. Hayashi and T. Nakagawa. Transient behaviors in the wien bridge
          oscillator: A laboratory experiment for the undergraduate student.
          Am. J. Phys., **52**(11):1021–1024, November 1984. The equation for
          the Wien bridge was found in this article.

[Mal93]   A.P. Malvino. *Electronic Principles*. Macmillan/McGraw-Hill, New
          York, New York, fifth edition, February 1993. pages 794–798.

[Moo92]   Francis C. Moon. *Chaotic and Fractal Dynamics*. Wiley, New York,
          NY, first edition, 1992. An excellent book, with some very nice ideas
          for experiments.

[Pip87]   A.P. Pippard. *Response and Stability*. Cambridge, Cambridge, Eng-
          land, first edition, 1987. A very interesting book. It contains many
          ideas for other simple experiments.